# Universal Design for Learning Science

## Reframing Elementary Instruction in Physical Science

# Universal Design for Learning Science

Reframing Elementary
Instruction in Physical Science

Deborah Hanuscin and
Delinda van Garderen

nsta Press
National Science Teaching Association
Arlington, Virginia

Claire Reinburg, Director
Rachel Ledbetter, Managing Editor
Jennifer Merrill, Associate Editor
Andrea Silen, Associate Editor
Donna Yudkin, Book Acquisitions Manager

**ART AND DESIGN**
Will Thomas Jr., Director, cover
Capital Communications LLC, interior design

**PRINTING AND PRODUCTION**
Catherine Lorrain, Director

**NATIONAL SCIENCE TEACHING ASSOCIATION**
1840 Wilson Blvd., Arlington, VA 22201
*www.nsta.org/store*
For customer service inquiries, please call 800-277-5300.

*NSTA is committed to publishing material that promotes the best in inquiry-based science education. However, conditions of actual use may vary, and the safety procedures and practices described in this book are intended to serve only as a guide. Additional precautionary measures may be required. NSTA and the authors do not warrant or represent that the procedures and practices in this book meet any safety code or standard of federal, state, or local regulations. NSTA and the authors disclaim any liability for personal injury or damage to property arising out of or relating to the use of this book, including any of the recommendations, instructions, or materials contained therein.*

**PERMISSIONS**
Book purchasers may photocopy, print, or e-mail up to five copies of an NSTA book chapter for personal use only; this does not include display or promotional use. Elementary, middle, and high school teachers may reproduce forms, sample documents, and single NSTA book chapters needed for classroom or noncommercial, professional-development use only. E-book buyers may download files to multiple personal devices but are prohibited from posting the files to third-party servers or websites, or from passing files to non-buyers. For additional permission to photocopy or use material electronically from this NSTA Press book, please contact the Copyright Clearance Center (CCC) (*www.copyright.com*; 978-750-8400). Please access *www.nsta.org/permissions* for further information about NSTA's rights and permissions policies.

**Library of Congress Cataloging-in-Publication Data**
Names: Hanuscin, Deborah L., author.
Title: Universal design for learning science : reframing elementary instruction in physical science /
    Deborah Hanuscin, Delinda van Garderen.
Description: Arlington, VA : National Science Teaching Association, [2020] | Includes bibliographical references
    and index. | Identifiers: LCCN 2019054086 (print) | LCCN 2019054087 (ebook) |
    ISBN 9781681406954 (paperback) | ISBN 9781681406961 (pdf)
Subjects: LCSH: Science--Study and teaching (Elementary)
Classification: LCC LB1585 .H295 2020  (print) | LCC LB1585  (ebook) | DDC 372.35/044--dc23
LC record available at *https://lccn.loc.gov/2019054086*
LC ebook record available at *https://lccn.loc.gov/2019054087*

# Contents

----

## Part I: The Frameworks

## Part II: The Vignettes

# Contents

## Part III: Applying the Frameworks

# Foreword

I was thrilled when Debi and Delinda asked me to write the foreword for this book—almost as thrilled as I was six years ago when they invited me to join their advisory board for the Quality Elementary Science Teaching (QuEST) program, upon which this book is based! I knew of Debi and Delinda's meaningful work in inclusive science education and was piqued by the prospect of a professional development program that used Universal Design for Learning (UDL) and the 5E Learning Cycle as the framework for quality science opportunities for all students. I am happy to share that my experiences on the advisory board only confirmed and strengthened my belief in the power and usefulness of this framework for the entire science education field. I also (fortunately!) came to know Debi and Delinda personally and found them to be incredibly warm, insightful, knowledgeable, and passionate educators who together form a "dynamic duo" in the inclusive STEM education community. This book is the fitting culmination of their brilliant project's work.

*Universal Design for Learning Science: Reframing Elementary Instruction in Physical Science* is a unique classroom resource, first and foremost, because of its assumption of *ability* rather than *disability*. Through highly readable text describing actual classroom experiences, this book dispels the myth that accessible science means "lowering the bar." Quite the contrary, the authors show how UDL in the context of thoughtfully crafted lessons aligned with the *Next Generation Science Standards* can provide appropriately challenging and dynamic experiences for *all* students, including those with special needs. Nine vignettes provide the reader with a window into lesson implementation in diverse science classrooms, while the accompanying "Teaching Tip" boxes provide easy-to-implement UDL strategies that reduce or eliminate unnecessary barriers to learning. Each of these tips can easily be incorporated across the curriculum!

Another strength of this book lies in its use of teacher-authors' voices. Teacher-participants from the QuEST program wrote most of the vignettes, so the reader is assured of real-world, teacher-tested strategies. Teacher educators will also find a chapter written specifically for them to support their work with methods classes. I personally found the background information on the

5E Learning Cycle and the crystal-clear instructions on developing coherent science storylines in Part I particularly helpful both for my own curriculum development and for instruction of my teacher candidates. Even as a veteran science educator, I experienced many "Aha" moments as I read this wonderful book!

I sincerely hope you share my belief that every child is entitled to quality science learning opportunities; in my mind, it is a practical, legal, and moral necessity. The field of science benefits from many diverse voices while society benefits from a scientifically literate citizenry. But what motivates me in my work (and perhaps you in yours) is the unyielding belief that all children deserve to experience the incredible, incomparable, breathtaking beauty and wonder of science! For that reason, I am most grateful for *Universal Design for Learning Science*, as I know it will inspire elementary school teachers to inspire their students—*all* their amazing students—in science.

—**Sami Kahn**
Executive Director
Council on Science and Technology
Princeton University

# Preface

*Deborah Hanuscin*

One of my most poignant memories of teaching elementary school took place on the playground. As I was watching a lively game of four square, I was joined on the sidelines by one of my students, Ricki. He had begun the year with me, transferred to another school for several months, and then rejoined our classroom just that week.

"I'm really glad that I'm back, Ms. H. I like being in your class better than my other one," he said. "Why's that?" I asked, my eyes still on the game. "Well, I'm not stupid in your class, but I was in the other one." I was taken aback. "What do you mean?" I inquired. "Well, I can do things in your class!" Ricki smiled, then ran over to take his turn at four square.

I was overwhelmed in that moment—not with joy over my apparent success as a teacher, but with renewed comprehension of how much the experiences of students in a classroom affect their self-perception. Here was a student—the same student—in two different classrooms, arriving at two drastically different conclusions about his abilities as a learner. While it might be tempting to conclude that I was a better teacher, I don't doubt that his other teacher was doing her best for Ricki to be successful. So, what were we doing differently?

Ricki was one of several students in my class with an individual education plan. I was helped in the implementation of this plan by a wonderfully talented special education teacher, Mrs. Branch, who insisted that Ricki could meet the same learning goals as other students, but that we needed to find alternative ways to support him in doing so. For example, if students were planning a science investigation, rather than having Ricki struggle to write out his plan in step-by-step detail, he was able to orally explain his ideas to Mrs. Branch, then fill in a graphic organizer summarizing the process he would use.

When creating a graph of his data to help with analyzing patterns, Ricki was able to use a calculator to find the average, rather than working out the computations by hand. When he needed to find information related to his investigation, Mrs. Branch would preselect nonfiction books at Ricki's reading level, and we would have students partner-read and share information they found in their books in small groups. Working with Mrs. Branch, I learned that the same learning activity could be implemented in a variety of ways, and that if one way posed a barrier for a student, another might allow him or her to be successful.

Given that we taught in the same school district, Ricki's other teacher was using the same curriculum materials as I was—curriculum materials that (unintentionally) were posing barriers to his learning. My best guess is that Mrs. Branch and I were more adept at recognizing these barriers and coming up with solutions. Because we held Ricki to the same standards as the other students, rather than "watering down" the learning goals, he felt just as capable as his peers when he was able to meet those goals. Ricki's story is not unique, though. Over the years I had many students who struggled—fourth graders who read at primer level or were still using invented spelling, students with dyslexia, kids with physical disabilities or emotional/behavioral disorders, and so on. What I noticed, however, is that despite their struggles, they experienced success in—and were therefore highly motivated for—learning science. Most of all, I found that adapting the activities to reduce barriers for these students did not lessen the challenge or rigor of the lessons—the students were succeeding at a level similar to their peers.

Though I certainly had a passion for teaching other subjects, I chose to specialize in science education during my graduate studies. I am now a teacher educator at a university. As an elementary school teacher, collaborating with a special educator enabled me to help all students be successful in science—and so, when I became a faculty member, I also sought out a collaborator in special education, my coauthor Delinda, who is also a former classroom teacher. Just as I learned from Mrs. Branch, I have learned a great deal from Delinda about meeting the diverse interests, needs, and abilities of students. (I think she's learned a little science from me along the way, too!) Our partnership, which began at the University of Missouri over a decade ago, has involved working with a variety of schools and teachers (and their students!) to promote "science for all" through the Quality Elementary Science Teaching (QuEST) program (see *https://sites. google.com/view/sciencequestprogram*).

### Delinda van Garderen

As an elementary school teacher, I loved designing hands-on and engaging experiences for my learners. While science was not my favorite subject to teach, I tried to make sure that these experiences were motivating and fun. Though I ended up pursuing a career as a professor in special education, I've kept a strong focus on the content that teachers are helping students learn, and I'm particularly interested in struggling learners in mathematics. However, through my partnership with Debi on the QuEST project, I have expanded my knowledge of science.

Today, I no longer have my own elementary school classroom full of kids, but what I have are college students preparing to be K−12 special education teachers or teacher educators. Many of the ideas related to science teaching and Universal Design for Learning (UDL) have great applicability in my own instruction at the university level. Some of my favorite courses to teach use the 5E Learning Cycle as a way to create meaningful learning opportunities both face-to-face and online. One of my favorite assignments involves my special education students collaborating with a general education teacher (in science or social studies) to redesign a lesson using UDL to make it accessible for specific struggling learners in the classroom. My partnership with Debi has been a great model for this kind of collaboration.

# Acknowledgments

We would like to acknowledge the many individuals we have collaborated with over the years through the Quality Elementary Science Teaching (QuEST) program, including our external evaluators (Mark Ehlert, Kristine Chadwick, and Tracy Bousselot), who gave extremely helpful feedback to refine our approach and improve our effectiveness. Additionally, we owe a debt of gratitude (and many cups of coffee) to our staff members (Karen King, Jesse Kremenak, Zandra de Araujo, Dante Cisterna, Tracy Hager, Betsy O'Day, Cathy Thomas, Annie Arnone, Kelsey Lipsitz, Kate Sadler, Jessi Keenoy, and Sarah Hill), who contributed their time, effort, expertise, and insights.

We are especially grateful to our contributing authors, and to all the teacher-participants in the QuEST program who gave up their summers, opened their classrooms to us, and dedicated their efforts to supporting the success of all students in science. Most importantly, we are grateful to *you* for purchasing this book—we hope that you, like us, have also found a collaborator, critical friend, or partner in crime to join you on your science teaching and learning journey!

Finally, we would like to acknowledge that this material is based upon work supported by the National Science Foundation under Grant No. DRL-1316683. Any opinions, findings, and conclusions, or recommendations expressed in this book are those of the authors and do not necessarily reflect the views of the National Science Foundation.

# About the Authors and Contributors

## Authors

**Deborah Hanuscin** is an experienced elementary classroom teacher and informal science educator, and since becoming a researcher, she has continued to find ways each year to teach elementary students. She received her PhD in science education from Indiana University and worked at the University of Missouri before moving to her current home in science, math, and technology education and elementary education at Western Washington University. She has led both state and federally funded projects working with teachers as well as students, and she has published more than 60 articles and book chapters for researchers, teachers, and administrators.

Debi's accomplishments include teaching awards at the campus, district, state, and national levels, and most notably she was named Outstanding Science Educator in 2014 by the Association for Science Teacher Education. She is the author of numerous articles in National Science Teaching Association (NSTA) journals and has served on the NSTA Research Committee, the NSTA Alliance of Affiliates, and the advisory boards of *Connected Science Learning* and *NSTA Reports*.

**Delinda van Garderen** is an experienced elementary and middle school classroom teacher in schools in New Zealand. Motivated by the questions she had about her own students and their diverse needs, she has focused her research in special education on struggling learners in science and mathematics.

Delinda received her PhD in special education from the University of Miami (Florida) and worked at State University of New York at New Paltz before accepting her current position at the University of Missouri in the Department of Special Education. She has led and been involved in both state and federally funded projects working with teachers and students in math and science, and she has published more than 40 articles and book chapters for researchers and teachers.

## *Contributors*

**Linda Buchanan** is a retired fourth-grade teacher from the Normandy School Collaborative. Her goal as a teacher was to empower students with the desire to become critical thinkers and life-long learners.

**Nicole Burks** is currently a fifth-grade teacher with the Columbia Public Schools District in Missouri. She loves integrating science throughout the day.

**Dante Cisterna** was a Postdoctoral Fellow on the Quality Elementary Science Teaching (QuEST) program. He taught science and biology for six years in middle and high school levels. Currently, he conducts research and assessment design in science education.

**Tracy Hager** is a third-grade teacher in Columbia Public Schools in Columbia, Missouri. She is a Presidential Awardee and National Board Certified Teacher, and she served as the elementary director on the Board of Directors of the Science Teachers of Missouri, an NSTA affiliate. She was also a contributing author to *Seamless Assessment in Science*.

**Kelsey Lipsitz** is a former elementary classroom teacher in the Hazelwood School District in Missouri. She received her PhD in science education from the University of Missouri, where she worked as a research assistant on the QuEST project. She is now a science educator with the Institute for Inquiry at the Exploratorium in San Francisco.

**Christine Meredith** is a fifth-grade teacher in the Hannibal School District in Hannibal, Missouri. She is part of the "science cohort" of teacher-leaders in her school.

**Betsy O'Day** teaches in the Hallsville School District in Hallsville, Missouri. She is a past president of the Science Teachers of Missouri, an NSTA affiliate. Betsy was part of the *Next Generation Science Standards* (NGSS) writing team, and she serves as an NSTA *NGSS* curator.

**Brooks Ragar** is a fifth-grade teacher in the Hannibal School District in Hannibal, Missouri. Following her participation in QuEST, she helped start a "science cohort" of teacher-leaders at Veterans Elementary School.

**Kate M. Sadler** worked as a special education teacher in Saint Louis, Missouri, for 15 years. She received her PhD in special education from the University of Missouri, where she worked as a research assistant on the QuEST project. She is currently a postdoctoral research associate for the Supporting Transformative Autism Research grant at the University of Virginia.

**Cody Sanders** was a preservice teacher when he attended the QuEST program through the University of Missouri. He is currently a fifth-grade teacher in Kansas City, Missouri, where he continues to incorporate the 5E Framework and Universal Design for Learning.

**Warren Soper** is retired after 21 years of teaching fourth grade in the Reeds Spring School District. He has been active in both NSTA and Science Teachers of Missouri (an NSTA affiliate), and he has presented at both their conferences.

**Mahaley Sullivan** is currently a fifth-grade teacher in the North Callaway School District in Williamsburg, Missouri. She has participated in QuEST twice and is a member of NSTA.

**Shelli Thelen** is a fifth-grade teacher at the Columbia Public School District in Missouri. After teaching 13 years in kindergarten, she made the jump up to fifth grade and has loved teaching science for the past four years. You can follow Shelli's classroom on Twitter (*@thelensthinkers*) or peek at her classroom blog at *www.thelensthinkers.blogspot.com*.

**Cathy Newman Thomas** was an associate professor at the University of Missouri when she worked with QuEST. She is now an assistant professor at Texas State University. Her interests are professional development and access to the general curriculum for diverse learners.

# Introduction

We are grateful to you for picking up our book! As former teachers, we understand that your time is precious and that you have many choices for how to spend it. This book is intended to honor that, and to provide you with tools and resources that can help you maximize the time you have to plan science lessons as well as engage students in learning science.

This book is the result of more than a decade of work with teachers through the Quality Elementary Science Teaching professional development program. In this work, we used two frameworks that come together in powerful ways to support student learning in science—the 5E Learning Cycle and Universal Design for Learning (UDL). Using these frameworks encourages teachers to rethink how they have typically approached lessons, and to *reframe* them in ways that mirror how students learn, that provide depth and conceptual coherence, and that support the success of all learners.

Implementing these frameworks doesn't require adopting a new curriculum (after all, we know well that teachers already have enough on their plates). Rather, it involves working with *existing curricula and resources* to identify barriers to learning and possible solutions. In other words, it means using a sharper knife, a bigger fork, or a deeper spoon to more effectively deal with what's already on your plate!

The information in this book will be useful to individual teachers seeking to improve their craft, or to groups of teachers collaborating to support student success in science. In particular, general educators and special educators who are coteaching science may find valuable common ground in the ideas presented in this volume. Even if you are familiar with these frameworks, we believe you will find something new within these pages.

## Part I: The Frameworks

Part I of the book provides an in-depth examination of the 5E Learning Cycle and UDL frameworks. We synthesize research that supports these frameworks and highlight common stumbling blocks to implementing them with success. For example, we emphasize the importance of including coherent conceptual storylines within a learning cycle sequence of activities and making assessment a "seamless" part of instruction. Embedded throughout this part are opportunities for you to "Stop and Consider" what you are reading and how the information aligns with your current ideas and practices for science teaching. For those already familiar with these frameworks, we will push you to deepen your understanding and self-evaluate what you know about how these can be implemented.

While Part I of the book will introduce you to the frameworks, we know that reading about them isn't enough for you to be able to implement them with success. The teachers in our program have reiterated the power of collaboration and the benefits of both experiencing new teaching approaches, as learners, and seeing them implemented by other teachers. Part II of this book provides examples of the successful implementation of these frameworks to teach physical science in elementary school classrooms. Rather than reading cover to cover, you may find it helpful to move back and forth between Part I and Part II to explore the examples and better understand these frameworks in action.

## Part II: The Vignettes

The teacher-authors we worked with in Part II of this book have done their best to provide you with a detailed view into both their classrooms and their instructional decision making. Each vignette begins with an overview of the lesson and the conceptual storyline that builds throughout the 5E Learning Cycle. Personal commentary from the teacher-authors provides additional insights into their teaching context and background and how they approached the lesson design. Embedded throughout are "Teaching Tip" boxes and "UDL Connection" listings to help make the teachers' thinking and design intentions explicit. The chapters conclude with a section that further unpacks teachers' application of UDL in terms of meeting the needs of specific learners and the lesson's alignment with the *Next Generation Science Standards* (NGSS).

The lesson vignettes featured in Part II highlight a variety of physical science topics that were the focus of our professional development program. Because our program took place during the transition to the *NGSS*, our teachers' stories are reflective of their own work to better align their instruction with the new standards. There are multiple possible routes to take to support students in meeting the performance expectations in the *NGSS*, and each vignette represents but a single route. The starting point for our teachers is often finding out what students know—whether by asking them to evaluate a claim, predict an outcome, or attempt to answer a question or explain a phenomenon. While these vignettes may or may not reflect the route you might take, we believe there is value in accompanying teachers on their journey.

We acknowledge that our decision to focus on individual lessons, as opposed to entire units, comes with some trade-offs. While the lessons are aligned to particular *NGSS* performance expectations (PEs), these may represent only a portion of the instructional activities necessary to support students in meeting those PEs. We chose to focus on single lessons to provide a more detailed picture of instruction using the 5E and UDL frameworks. Where possible, we've tried to incorporate contextual information to aid the reader in understanding the larger unit; however, we recognize that some aspects of the teaching and learning picture may not be visible to readers without the entire set of lessons. Additionally, we acknowledge it is unlikely that students will develop an understanding of the *NGSS* crosscutting concepts (CCCs) within a single lesson; rather, CCCs are themes that connect learning across topics and science disciplines. For this reason, you will find that many of the summative assessments included in the lessons do not address this third dimension explicitly.

Similarly, a hallmark of *NGSS*-aligned instruction is the focus on figuring out phenomena.[1] This can be accomplished in different ways—and not all phenomena need to be used for the same amount of instructional time. An anchoring phenomenon might serve as the overall focus for a unit, along with other investigative phenomena along the way as the focus of an instructional sequence or lesson. Lessons may also highlight everyday phenomena that relate investigative or anchoring phenomena to personally experienced situations. Within each vignette, we emphasize the investigative or lesson-level phenomenon, as well as point out the relevant anchoring phenomenon, where appropriate, that frames the overall unit.

The grade 3−5 classrooms featured in Part II also highlight a diversity of learners, and as such, particular classrooms may be of more interest to you given the topics you teach and the diversity of learners within your own classroom. Not all aspects of "diversity" are covered in this book, but the solutions presented may apply to other types of diverse learners you work with. Further, while each teacher applied UDL to meet the needs of *all students*, we have focused more deeply on highlighting several particular learners in each case to illustrate specific ways that the teacher applied UDL to meet those students' needs. The solutions applied represent what the teachers actually did in their classrooms.

---

1. See "Using Phenomena in NGSS-Designed Lessons and Units," *www.nextgenscience.org/sites/default/files/Using%20Phenomena%20in%20NGSS.pdf.*

Therefore, it may be possible that we have overlooked viable UDL connections or you may not agree with the UDL connections that were made. Keep in mind that these chapters serve as examples that we hope will resonate with you and will help you envision how you might undertake similar approaches with the 5E Learning Cycle and UDL in your own classroom.

## Part III: Applying the Frameworks

Part III of this book will provide you with additional resources to get you started reframing your instruction or using the book to support others in doing so. We highlight tools that we have used with teachers, as well as ways that we have integrated these frameworks into preservice teacher education courses—both in special education and elementary science education. We hope you will find ways of your own to make this book a useful part of your professional development or that of others, and we invite you to share with us what you do!

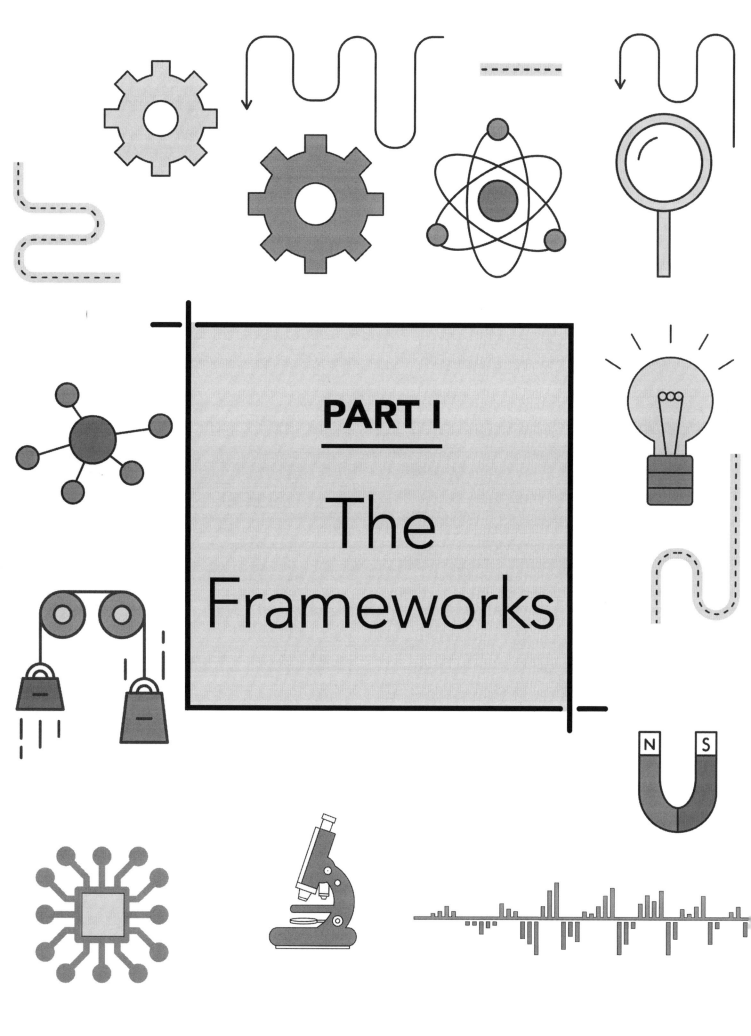

# PART I

## The Frameworks

# CHAPTER 1

# Reframing Instruction
# With the 5E Learning Cycle

**Dante Cisterna, Deborah Hanuscin, and Kelsey Lipsitz**

If you're interested in strengthening your understanding of the learning cycle to (re)frame your classroom instruction, we are here to help. You may already be familiar with the learning cycle—or at least some version of it. Since Robert Karplus and Herbert Thier first described it in 1967 for the Science Curriculum Improvement Study, the learning cycle now has several iterations, ranging from three to seven phases. Perhaps the most popular is the 5E Learning Cycle (Bybee 1997), which has been adopted and covered in several National Science Teaching Association publications including the *Picture Perfect Science* series and in many *Science and Children* articles. Yet, even among teachers who are familiar with the learning cycle, we find there are several common misconceptions that result in a less effective application of this instructional framework.

In this chapter, we provide an overview of the 5E (Engage, Explore, Explain, Extend, Evaluate) Learning Cycle and the research that supports the efficacy of this framework in science classrooms. We clarify how this learning cycle differs from instruction that is more traditional, and we discuss several points of confusion that we've encountered among teachers. We also outline additional frameworks that complement the learning cycle and enhance its application. For example, the idea of "seamless assessment" (Abell and Volkmann 2006) further elaborates on the importance of both formative and summative assessment throughout the 5E Learning Cycle. In addition, our own work surrounding conceptual storylines (Hanuscin et al. 2016) emphasizes the importance of building toward "big ideas" and conceptual coherence when implementing the learning cycle. Reframing your instruction using these tools will allow you to plan lessons that are content rigorous, student led, and meaningful for science conceptual development.

---

### STOP AND CONSIDER ...

Before you read the rest of this chapter, take a moment to examine three different versions of a lesson that might occur after students have already learned how to build a complete circuit using batteries, bulbs, and wires (see the table below). The lesson builds toward *Next Generation Science Standards* (*NGSS*) performance expectation 4-PS3-4: "Apply scientific ideas to design, test, and refine a device that converts energy from one form to another" (NGSS Lead States 2013).

| THREE VERSIONS OF A LESSON | | |
| --- | --- | --- |
| **Lesson 1** | **Lesson 2** | **Lesson 3** |
| **a.** Students are given two batteries, two bulbs, and several wires, and they are challenged to design a circuit in which the bulbs are the brightest. Afterward, the teacher has students share how they designed their circuit to accomplish the challenge. | **a.** Students are given a switch and are challenged to find a way to connect the switch to the circuit they have built to make the bulb go off and on. Afterward, the teacher has students share how they accomplished the challenge and describe how the switch affects the function of the circuit. | **a.** Students are introduced to a switch and how it works in a reading in their science textbook. |
| **b.** The teacher distributes different kinds of light bulbs (e.g., spotlights, holiday bulbs, compact fluorescents) for students to examine. Students compare the different parts of each bulb (e.g., filament, screw threads) and how bright it glows when turned on. | **b.** The teacher distributes examples of different kinds of household switches (from lamps, stereos, refrigerator lights, etc.) for students to examine. Students compare the parts and function of different switches and their purpose (e.g., "off" when depressed for a refrigerator light versus off and on for a lamp). | **b.** The teacher distributes switches to students and together they make a labeled diagram of the switch, showing the important connection points. |
| **c.** The teacher facilitates a whole-class discussion in which students generate a list of the parts of a light bulb and the role they play in the function of the bulb (e.g., the screw threads are a contact point but also help secure the bulb in place). | **c.** The teacher facilitates a whole-class discussion in which students generate a list of important design considerations for switches so that they work as intended (e.g., the part you touch should be an insulator, how two conducting parts should connect, where the switch should be placed in the circuit). | **c.** The teacher distributes additional circuit materials and asks students to integrate the switch into a circuit as illustrated so that a bulb can be turned off and on. |

*(continued)*

---

| THREE VERSIONS OF A LESSON *(continued)* | | |
|---|---|---|
| **Lesson 1** | **Lesson 2** | **Lesson 3** |
| **d.** The class is provided an example of a circuit in which the bulb is not lighting, and students are challenged in small groups to figure out how they could correct the problem (i.e., determine whether the problem is a burnt-out bulb or something else). | **d.** The class is provided an example of a circuit in which the switch is not working properly, and students work in small groups to troubleshoot and correct the problem (e.g., correct the connections to the circuit). | **d.** The class is provided an example of a circuit in which the switch is not working properly, and students work in small groups to troubleshoot and correct the problem (e.g., correct the connections to the circuit). |
| **e.** The teacher provides students with an assortment of materials (paper clips, foil, fasteners, card stock, etc.) and asks them to design a signal lamp that could be used as a backup device in case of a complete failure of an aircraft's radio. Students demonstrate their devices and explain how they work. | **e.** The teacher provides students with an assortment of materials (paper clips, foil, fasteners, card stock, etc.) and asks them to design a signal lamp that could be used as a backup device in case of a complete failure of an aircraft's radio. Students demonstrate their devices and explain how they work. | **e.** The teacher provides students with instructions for making a signal lamp that could be used as a backup device in case of a complete failure of an aircraft's radio. Using a simple code provided by the teacher, students demonstrate their work by sending a message to their classmates. |

Take a moment to jot down your ideas about the three lessons. How are they alike? Different? Which do you feel is the strongest lesson? Why?

## What Does Research Say About the Learning Cycle?

There is increasing pressure on teachers and schools to use "evidence-based" practices, yet few teachers have subscriptions to science education research journals. Lucky for you, we have access to research and have identified some key findings with regard to the learning cycle and the evidence for its effectiveness. In fact, use of the learning cycle is perhaps one of the most well-evidenced practices in science education over the past five decades![1] Over the years, multiple studies emphasize that the learning cycle model helps students make sense of scientific ideas, improve their scientific reasoning, and increase their engagement in science class (Lawson 1995). Research shows that when exploration precedes concept introduction, as in the learning cycle, students exhibit greater achievement and retention of concepts (Abraham 1998; Renner, Abraham, and

---

1. In 2006, the Biological Sciences Curriculum Study (BSCS) issued a report on the origins and effectiveness of the 5E Learning Cycle (available at *https://bscs.org/sites/default/files/_media/about/downloads/BSCS_5E_Full_Report.pdf*).

Birnie 1988), particularly in comparison to other instructional models (Akar 2005; Bishop 1980; Bowyer 1976; Nussbaum 1979; Renner and Paske 1977; Saunders and Shepardson 1987; Schneider and Renner 1980).

The 5E Learning Cycle (Bybee 1997) is an instructional model consistent with a synthesis of research on effective science instruction (Banilower et al. 2010), and it fits with the vision of the *NGSS* (NGSS Lead States 2013).[2] However, we caution that the efficacy of the learning cycle depends on how well teachers implement the model with fidelity. For example, Coulson (2002) found that the learning gains of students whose teachers had medium or high levels of fidelity to the 5E Instructional Model were nearly double that of students whose teachers did not use the model or used it with a low level of fidelity. If you want to be sure that your students are getting the most out of your implementation of the learning cycle, read on!

## How Does the Learning Cycle Differ From Traditional Instruction?

As shown in Table 1.1, there are some key differences between what might be referred to as instruction that is more "traditional" and instruction using the 5E Learning Cycle, particularly in terms of the roles and activities of the teachers and students during the lesson.

| TABLE 1.1. The Learning Cycle vs. Traditional Instruction | | | | |
|---|---|---|---|---|
| | The Learning Cycle | | Traditional Instruction Sequence | |
| Phase of Lesson | *Activities of the Teacher* | *Activities of the Students* | *Activities of the Teacher* | *Activities of the Students* |
| Engage | Elicit student ideas; establish a context for the lesson that motivates and engages students | Self-evaluate what they know about the topic | Provide students with background information and vocabulary | Listen carefully to new information that is presented |
| Explore | Provide experience with a phenomenon | Test their existing ideas about evidence | Provide hands-on activities to reinforce the concepts | Verify new information by conducting an experiment or test |

*(continued)*

---

2. See STEM Teaching Tools Practice Brief 4, "Are there multiple instructional models that fit with the science and engineering practices in *NGSS*? (Short answer: Yes)," at *http://stemteachingtools. org/brief/4*.

| | The Learning Cycle | | Traditional Instruction Sequence | |
| Phase of Lesson | Activities of the Teacher | Activities of the Students | Activities of the Teacher | Activities of the Students |
|---|---|---|---|---|
| **TABLE 1.1. The Learning Cycle vs. Traditional Instruction (*continued*)** | | | | |
| Explain | Support student sense-making and introduce academic language | Make sense of new evidence and build new explanations | Clarify and further explain the concepts to students | Comprehend the explanation provided by the teacher or text |
| Extend | Provide opportunities for students to apply new ideas across a variety of contexts | Apply new understandings to a different context | If time is left, provide enrichment or extension activities | If opportunity is provided, apply the information to a new context |
| Evaluate | Summatively assess students' learning and identify next steps for instruction | Demonstrate new understandings and new academic vocabulary | Evaluate students (e.g., test) | Regurgitate the information introduced in the lesson |

Traditional instruction usually starts by introducing scientific concepts and vocabulary, and the subsequent activities aim for verification of the concepts—for example, through hands-on activities and investigations. The lesson is highly teacher centered, with the teacher assuming most of the responsibility for providing explanations. By contrast, in the 5E Learning Cycle, students have the opportunity to make their initial ideas explicit and test their ideas through exploration activities, using evidence to develop scientific explanations for themselves. This is a more student-centered approach, in which the teacher assumes the role of facilitator—asking questions, probing student ideas, and guiding discussions. While this may sound appealing, switching from a more traditional approach to a learning cycle approach can be difficult. In the sections that follow, we share some of the difficulties that teachers encounter—and how you can move toward effective implementation of the learning cycle in your own teaching.

### *Everything Must Go!*

One of the first challenges in implementing the 5E Learning Cycle is the fear that everything you've been doing all along must be discarded—that you have to "throw the baby out with the bathwater," so to speak. In the Quality Elementary Science Teaching (QuEST) professional development program, teacher-participants came from different schools and districts—and they often had different curriculum materials they were expected to implement. For those reasons, we were interested in helping our teachers develop knowledge and skills that could be used and adapted to their own school and classroom contexts—regardless of the specific curriculum materials used. Like our teachers, you will be able to apply the learning cycle to your existing lessons and units. You won't have to throw out what you are already doing in its entirety.

The 5E Learning Cycle can be used as a framework to help you sequence your instruction and leverage your curriculum materials in more powerful ways. For example, if you have a textbook, the learning cycle suggests that having students read informational texts might best occur in the Explain phase, to help the students make sense of their experiences by connecting their ideas to scientific explanations. Or you might notice that the activity in your kit may be better introduced as an exploration after adding a formative assessment activity to engage students and elicit their ideas. Finally, you may notice that your curriculum suggests some optional lesson extensions "if time allows"—you might realize that rather than being add-ons, these are essential opportunities to allow students to apply their new ideas and understanding, and thus they should definitely be included in your lesson.

### *Lesson vs. Session*

We've found that the term *lesson* is often used synonymously with one class session. When viewed this way, the idea of implementing a learning cycle lesson in one class period can seem like a daunting task! Given that teachers have different lesson schedules and amounts of instructional time for science, we find that thinking about a *lesson* separately from a *class session* can be helpful. That is, a learning cycle may be considered a single lesson that encompasses a series of activities (sequenced in several phases) to help students develop an understanding of a particular science idea. These activities might be accomplished in one teaching session, but more likely they will be spread out over several days. In particular, it is the order of the activities that is especially important when implementing the learning cycle.

### *Explore First, Explain Later*

Regardless of the version of the learning cycle used, a feature common to all is the fact that exploration precedes explanation. This makes sense, particularly in light of the *NGSS* practices of constructing explanations (Science and Engineering Practice [SEP] 6) and engaging in argument from evidence (SEP 7). The exploration phase activities help students gather the necessary evidence with which they will build their explanations. We find this to be counterintuitive, however, for teachers who are used to introducing information—for example, with a textbook reading—prior to students completing activities. With preservice teachers in particular, we find that they believe they need to provide students with "background information" before beginning an activity (see also Otero and Nathan 2008). Rather than finding out what background and prior knowledge students can bring to bear on the activity, teachers essentially "give away the punchline" by explaining the concepts to the students before they participate in experiences that might have led them to these same ideas. This relates to another point of confusion about the learning cycle that we've encountered—specifically, the purpose of the Engage phase.

### Vocabulary First!

You may be used to "frontloading" vocabulary at the start of a lesson—similar to providing background information as discussed above. Vocabulary is indeed important to science learning—for without becoming proficient with the academic language of science, students cannot readily engage in the type of deep learning that will enable them to go beyond memorizing facts (NASEM 2018). However, the learning cycle places vocabulary instruction in the Explain phase— after students have concrete and firsthand experiences to which they can attach new terms and build proficiency with using them.

### Engagement Is Having Fun

The very use of the term *engagement* seems to connote building excitement in the lesson. This is most certainly important—but not in terms of keeping students entertained or having fun. Rather, it extends to students' emotional investment in learning:

> *Emotion plays a role in developing the neural substrate for learning by helping people attend to, evaluate, and react to stimuli, situations, and happenings. … Quite literally, it is neurobiologically impossible to think deeply about or remember information about which one has had no emotion because the healthy brain does not waste energy processing information that does not matter. … People are willing to work harder to learn the content and skills they are emotional about, and they are emotionally interested when the content and skills they are learning seem useful and connected to their motivations and future goals.* (NASEM 2018, pp. 29–30)

Yet, engagement also refers to *intellectually engaging* students in the phenomenon or question of focus. This first phase of the learning cycle also acknowledges what we know about how people learn, namely:

_Students come to the classroom with preconceptions about how the world works. If their initial understanding is not engaged, they may fail to grasp new concepts and information that are taught, or they may learn for the purpose of a test but revert to their preconceptions outside the classroom._ (Bransford, Brown, and Cocking 1999, p. 10)

Thus, eliciting students' ideas and tapping into their funds of knowledge is a necessary step toward successful implementation of the learning cycle. It's not enough, however, to merely elicit student ideas—assessment becomes formative only when it is used to inform instruction. Being purposeful in assessment throughout the learning cycle is important. Later in the chapter, we discuss how the concept of "seamless assessment" (Abell and Volkmann 2006) can help you accomplish that.

### Who Is Doing the Explaining?

Just as finding out students' ideas _prior to_ engaging in activities is important, allowing students to articulate their developing ideas _following_ participation in activity is essential. In the Explain phase of the lesson, however, we find it is tempting for teachers to do all of the explaining themselves—leaving little room for students to grapple with ideas and make sense of their experiences. For this reason, we find it helpful to foreground student explanation in this phase of the lesson. The teacher plays an important role in facilitating the sense-making but shouldn't do the thinking for the students. Of course, there are limitations to what might be possible for students to explain based on their explorations, as we consider below.

### Students Learn From Hands-On Activity

A final challenge we encounter with teachers relates to the notion that students can learn everything through doing a hands-on activity. It stands to reason that while firsthand experiences with materials are certainly important, other learning activities such as participating in role-play, engaging with children's literature, and using computer simulations should not be overlooked. Yet, more problematic is the assumption that by simply manipulating materials, students will come to an understanding of a concept. Take, for example, a student who uses a magnet and observes the phenomenon that some objects are attracted and some are not. He or she may realize that some metals, but not others, fit into the first group of objects that are attracted, but simply doing this activity is insufficient to help the student understand _which metals_ interact with magnets in this way, or why they do so.

That is, the idea that magnets attract only iron, nickel, and cobalt would not be possible for students to conclude following this kind of exploration. Without comprehending this idea, it could be difficult for students to select the appropriate materials when designing a way to use magnets to solve a problem. Understanding the strengths and limitations of various activities in terms of the ideas students can develop is essential, as is providing appropriate time and support for student sense-making. While we emphasized the importance of students doing the explaining in the previous section, there are indeed times when teachers must scaffold those explanations with additional information—whether they provide it themselves or it comes through students' further research into the phenomenon (such as consulting informational texts).

## STOP AND CONSIDER ...

Return to the notes you made about the three versions of the lesson at the beginning of this chapter. Which version(s) more closely resemble traditional instruction? Which version(s) resemble the 5E Learning Cycle? What new ideas do you have about using the learning cycle to frame instruction?

## Conceptual Storylines: Linking Ideas and Activities

The 5E Learning Cycle purports to organize classroom activities in a way that is meaningful for students. Let's illustrate this with an example. Read the following sequence of words:

- *A*
- *Activities*
- *Be*
- *Cycle*
- *Of*
- *In*
- *Learning*
- *Meaningful*
- *Must*
- *Sequenced*
- *The*
- *Ways*

You may notice that this list is sequenced in alphabetical order, which is a traditional method for organizing lists.

Now read this sequence of words:

*The activities of a learning cycle must be sequenced in meaningful ways.*

Notice a difference? Just as sentence structure helps convey the meaning of a series of words, the way in which we organize and sequence activities in a lesson can either obscure or help reveal meaning to students!

An effective learning cycle sequence of activities in a lesson should have a coherent *conceptual storyline*, which refers to the flow and sequencing of learning activities so that concepts align and progress in ways that are instructionally meaningful to student learning (Ramsey 1993). Although every learning cycle has a storyline, not every storyline is conceptually coherent. Sequencing and connecting scientific concepts in a coherent storyline is important because this structure can help provide meaning for students, and thus it affects their learning (Roth et al. 2011). Scientific concepts need to be organized so that they build on each other in ways that help students understand the lesson's big idea—much like a learning progression is used to sequence ideas across units and grade levels. While conceptual understanding is not the only learning goal of the lesson, we find conceptual storylines to be especially useful for elementary school teachers who typically have less preparation in science. In this manner, we acknowledge that they focus on only *one* of the dimensions emphasized in the *NGSS*—but one for which elementary teachers may need support.

Early in the QuEST program, we noticed that our participating teachers were able to align activities successfully to the purpose of each phase of the 5E Learning Cycle. For example, they chose formative assessment tasks for the Engage phase, picked hands-on activities for the Explore phase, and used appropriate types of application and synthesis activities for the Extend phase. Yet the activities themselves often did not connect in meaningful ways. Think back to our earlier sentence:

> *The activities of a learning cycle must be sequenced in meaningful ways.*

Now, compare that sentence to this one:

> *The dogs of a reading bicycle must be raised in beautiful approaches.*

Both have the same structure—the same kinds of words in each position in the sentence (e.g., nouns, adjectives)—but the words in the second sentence do not convey an overall meaning.

In our work with teachers, we found that the conceptual storyline of a lesson was an element largely invisible to them. Indeed, when we asked what makes a "good" science lesson, typical responses included the following:

- Lessons that are engaging and interesting to students
- Lessons that relate to students' everyday lives
- Lessons that are student centered

While these are all important to good instruction, activities should also help students develop conceptual understanding.

A Trends in International Mathematics and Science Study video study (Roth and Garnier 2006), which compared math and science instruction around the world, suggested that student achievement in the United States could be enhanced with renewed attention to the concepts that activities are intended to help students understand. Our own research (Hanuscin et al. 2016) illustrated that teachers often connect activities by *topic* as opposed to *concept*, which can result in conceptual breadth versus depth in lessons. Rather than trying to address multiple concepts and ideas at a surface level, a good learning cycle lesson will tackle a single concept or big idea in depth.

## STOP AND CONSIDER ...

Look back at our three versions of a lesson again, this time attending to the *concepts* underlying each of the activities. What do you notice? How do these ideas relate? Is there one concept or many? Does the lesson promote depth or breadth of understanding? Jot down some ideas before reading on.

## Putting It All Together

Let's revisit the three different versions of the lesson at the beginning of the chapter and consider the following:

- To what extent do the selected activities meet the purpose of each phase of the learning cycle?
- What is the key idea that each activity helps students understand?
- How do those ideas connect from activity to activity?

Both Lesson 1 and Lesson 2 begin with activities that allow the teachers to gain insights into students' prior knowledge about the topic (Engage phase), whereas Lesson 3 begins with the teacher introducing students to a switch and how it works through a reading in their science textbook (a feature of traditional instruction). Lessons 1 and 2 continue with a hands-on exploration (Explore), while Lesson 3 has students draw a labeled diagram of a switch with guidance from the teacher. The teachers in Lessons 1 and 2 then facilitate a whole-group discussion in which the students generate a list of the parts of a light bulb and the role the parts play, in the case of Lesson 1, and a list of important design considerations for switches, in the case of Lesson 2 (Explain). The teacher in Lesson 3, however, asks students to integrate the switch into a circuit as illustrated rather than participate in an activity that enables them to generate their own explanations.

All three lessons then engage students in a troubleshooting challenge, but in Lessons 1 and 2, the students are given the opportunity to connect their ideas to a new, real-world context (Extend) following a hands-on exploration and sense-making activity. The last activity, although very similar across all three lessons, differs in one critical way—in Lesson 3, the students are given instructions for making a signal lamp, whereas the students in Lessons 1 and 2 have the opportunity to communicate their new learning through their own design of a signal lamp (Evaluate).

Although both Lesson 1 and Lesson 2 follow the structure of the 5E Learning Cycle, this does not mean that they both have coherent conceptual storylines. The teacher in Lesson 1 begins the lesson by eliciting students' ideas about how to design a circuit in which the bulbs are the brightest, but this storyline is not developed further in the activities that follow. The focus of Lesson 1 switches (pun intended!) from the brightness of bulbs to the parts of a light bulb and the roles they play, to troubleshooting a circuit in which the bulb is not lighting, to using students' knowledge of switches to design a signal lamp. Lesson 2, however, begins with eliciting students' ideas about switches and continues this storyline by deepening an understanding of the role that switches play in a system throughout the 5E Learning Cycle. Although Lesson 3 follows a more traditional lesson structure, it also focuses on switches throughout the lesson. In summary, Lesson 2 is the only lesson that follows the 5E Learning Cycle *and* has a coherent conceptual storyline.

The conceptual storyline is often an implicit consideration during lesson plan design, so we worked to make it more explicit. In the QuEST program, we adapted and developed tools for helping teachers develop coherent conceptual storylines. For example, we adapted the conceptual storyline map from Bybee (2015) to make explicit the concepts that students are developing in each phase of the 5E Learning Cycle. We also included linking questions between the concepts developed in each phase to ensure that the concepts are connected and arranged in a coherent way (see the example of Figure 1.1).

We often find that students' spontaneous questions during one phase of the learning cycle relate to the activities and ideas we plan to focus on in the next phases—a sign that we have sequenced the activities appropriately. However, it's important to monitor the development of students' ideas throughout to understand the meaning that they are developing through the sense-making activities in each phase of the lesson.

**Figure 1.1.** Charting a Conceptual Storyline Map

# Conceptual Storyline Map

**Engage**
Challenge to integrate a switch into a circuit

**Key idea:** Switches can be used to open and close a circuit

**Linking question:** Do all switches work the same way?

**Explore**
Examining different kinds of switches

**Key idea:** Some switches work by pressing, others by flipping

**Linking question:** What do all switches have in common?

**4-PS3-4:**
Apply scientific ideas to design, test, and refine a device that converts energy from one form to another.

**Explain**
Discussing design considerations

**Key idea:** Switches are made of conductors and insulators

**Evaluate**
Engineering design task

**Key idea:** Switches can control the flow of energy in a circuit

**Linking question:** How can I use a switch in my device?

**Extend**
Trouble-shooting a circuit

**Key idea:** A change to one part of a system can affect the entire system

**Linking question:** What if a switch doesn't work as intended?

*Source:* Adapted from Bybee (2015).

## STOP AND CONSIDER ...

In the lesson discussed, what are the opportunities for assessment that occur in each phase of the learning cycle? What kinds of evidence of student understanding would a teacher be able to use at each point to determine whether the students were ready for the next phase of the lesson? Jot down some ideas before reading on.

## An Assessment Cycle: The 5E Model and Seamless Assessment

Multiple assessment opportunities throughout a learning cycle allow teachers to identify and build on students' existing ideas, collect evidence of developing student understanding, and make decisions based on that evidence. Similarly, students need to have evidence to self-assess their own knowledge and skills. In the QuEST program, we used Abell and Volkmann's (2006) model of "seamless assessment," which provides explicit strategies for embedding formative assessment within each phase of the 5E Learning Cycle to inform teaching and learning. In our professional development experience, we have seen that when teachers have a grounded understanding of the 5E Learning Cycle (the ability to select and sequence activities with purpose) and the conceptual storyline (to provide coherence in the scientific concepts), they can identify the key ideas to assess throughout the lesson and the strategies that are appropriate for collecting evidence of student learning.

Often, these strategies are embedded in the activities themselves, so they are not separate from instruction (see Table 1.2). For example, a "friendly talk probe" (Keeley 2008) feels more like a learning activity than a test, but it is a useful way to assess students' ideas and identify potential misconceptions. Additionally, these assessment strategies are connected to the main learning goal of the lesson, the key concepts as identified in the conceptual storyline, and the activities or content representations provided to students. This enables the teacher to monitor the development of students' ideas, to assess their progress and readiness to move on in the lesson, or to determine when changes in instruction need to be made to address student difficulties and better support student learning. The conceptual storyline depends on teachers' scientifically accurate conceptions, so they can compare students' ideas to these and use the storyline as a map to help them develop their ideas further.

| TABLE 1.2. Seamless Assessment Strategies Embedded in the Lesson | | |
|---|---|---|
| **Lesson 2** | **Key Ideas** | **Seamless Assessment Strategies** |
| **a.** Students are given a switch and are challenged to find a way to connect the switch to the circuit they have built to make the bulb go off and on.<br><br>Afterward, the teacher has students share how they accomplished the challenge and describe how the switch affects the function of the circuit. | Switches can be used to open and close a circuit, and they control the flow of energy. | Initial discussion and brainstorming: The teacher first asks students about their past experiences with switches and how they think they work. (What ideas do students already have about how switches work?)<br><br>Observation of students' work during the challenge. (How are they using what they have learned already about a complete circuit to integrate the switch?) |
| **b.** The teacher distributes examples of different kinds of household switches (from lamps, stereos, refrigerator lights, etc.) for students to examine. Students compare the parts and function of different switches and their purpose (e.g., "off" when depressed for a refrigerator light versus off and on for a lamp). | Some switches work by depressing a part, others by moving a part back and forth. | Annotated drawing: Students create a labeled drawing that explains the parts and function of a switch.<br><br>Box-T chart: Students use this graphic organizer to record things that all switches have in common and the differences between various types of switches (used in the next phase as a discussion tool). |
| **c.** The teacher facilitates a whole-class discussion in which students generate a list of important design considerations for switches so that they work as intended (e.g., the part you touch should be an insulator, how two conducting parts should connect, where the switch should be placed in the circuit). | All switches have two points (conductors) at which they connect to a circuit and another connection that can be opened or closed in different ways. The "switch" is a conductor to protect the user. | Science notebooking: Students create a list of design considerations as a reference tool for use later in the lesson. |

*(continued)*

**TABLE 1.2. Seamless Assessment Strategies Embedded in the Lesson (*continued*)**

| Lesson 2 | Key Ideas | Seamless Assessment Strategies |
|---|---|---|
| **d.** The class is then provided an example of a circuit in which the switch is not working properly, and students work in small groups to troubleshoot and correct the problem (e.g., correct the connections to the circuit). | One part of a system can affect the function of the entire system. The switch must be connected as part of a complete circuit to operate as intended. | Friendly talk probe in which different fictional students offer suggestions for how to fix the circuit. Students indicate with whom they most agree, and they justify their responses. (Are students able to apply their knowledge of the structure or function of a switch?) |
| **e.** The teacher then provides students with an assortment of materials (paper clips, foil, fasteners, card stock, etc.) and asks them to design a signal lamp that could be used as a backup device in case of a complete failure of an aircraft's radio. Students demonstrate their devices and explain how they work. | Switches can be used with a light to signal and communicate over a distance. | Formal student presentations and peer evaluation-critiques of initial designs with suggestions for improvement. (Are students applying their knowledge of switches and taking into account the design considerations? Can they identify potential failure points and ways to optimize their design?) |

## Next Steps

Even a well-designed learning cycle lesson with a coherent conceptual storyline may not be suitable for the diverse needs, interests, and abilities of your specific students. Specific activities, representations, or modes of instruction may pose barriers to some learners. The next chapter will introduce you to the Universal Design for Learning framework, which can complement the frameworks introduced in this chapter and help you identify barriers and possible solutions for the diverse learners in your classroom.

Rather than moving on to the next chapter, however, you may find it helpful at this point to explore some of the lesson vignettes and snapshots in Part II of this book to develop a stronger understanding of how the 5E Learning Cycle, coherent conceptual storylines, and seamless assessment are used to frame instruction. The choice is yours!

## References

Abell, S. K., and M. J. Volkmann. 2006. *Seamless assessment in science: A guide for elementary and middle school teachers*. Arlington, VA: NSTA Press.

Abraham, M. R. 1998. The learning cycle approach as a strategy for instruction in science. In *International handbook of science education*, ed. B. J. Fraser and K. G. Tobin, 513–524. Dordrecht, the Netherlands: Kluwer Academic Publishers.

Akar, E. 2005. Effectiveness of 5E learning cycle model on students' understanding of acid-base concepts. MS thesis, Middle East Technical University, Ankara, Turkey. *https://etd.lib.metu.edu.tr/upload/12605747/index.pdf*.

Banilower, E., K. Cohen, J. Pasley, and I. Weiss. 2010. *Effective science instruction: What does research tell us?* 2nd ed. Portsmouth, NH: RMC Research Corporation, Center on Instruction.

Bishop, J. E. 1980. The development and testing of a participatory planetarium unit employing projective astronomy concepts and utilizing the Karplus learning cycle, student model manipulation and student drawing with eighth-grade students. *Dissertation Abstracts International* 41 (3): 1010A.

Bowyer, J. A. B. 1976. Science Curriculum Improvement Study and the development of scientific literacy. *Dissertation Abstracts International* 37 (1): 107A.

Bransford, J. D., A. L. Brown, and R. R. Cocking. 1999. *How people learn: Mind, brain, experience, and school*. Washington, DC: National Research Council.

Bybee, R. W. 1997. *Achieving scientific literacy: From purposes to practices*. Portsmouth, NH: Heinemann.

Bybee, R. W. 2015. *The BSCS 5E instructional model: Creating teachable moments*. Arlington, VA: NSTA Press.

Coulson, D. 2002. *BSCS science: An inquiry approach—2002 evaluation findings*. Arnold, MD: PS International.

Hanuscin, D., K. Lipsitz, D. Cisterna-Alburquerque, K. A. Arnone, D. van Garderen, Z. de Araujo, and E. J. Lee. 2016. Developing coherent conceptual storylines: Two elementary challenges. *Journal of Science Teacher Education* 27 (4): 393–414.

Karplus, R., and H. D. Thier. 1967. *A new look at elementary school science*. Chicago: Rand McNally.

Keeley, P. 2008. *Science formative assessment: 75 practical strategies for linking assessment, instruction, and learning*. Thousand Oaks, CA: Corwin Press and NSTA Press.

Lawson, A. E. 1995. *Science teaching and the development of thinking*. Belmont, CA: Wadsworth Publishing Company.

National Academies of Sciences, Engineering, and Medicine (NASEM). 2018. *How people learn II: Learners, contexts, and cultures*. Washington, DC: National Academies Press. *https://doi.org/10.17226/24783*.

NGSS Lead States. 2013. *Next Generation Science Standards: For states, by states*. Washington, DC: National Academies Press. *www.nextgenscience.org/next-generation-science-standards*.

Nussbaum, J. 1979. Children's conceptions of the Earth as a cosmic body: A cross age study. *Science Education* 63 (1): 83–93.

Otero, V. K., and M. J. Nathan. 2008. Preservice elementary teachers' views of their students' prior knowledge of science. *Journal of Research in Science Teaching* 45 (4): 497–523.

Ramsey, J. 1993. Developing conceptual storylines with the learning cycle. *Journal of Elementary Science Education* 5 (2): 1–20.

Renner, J. W., M. R. Abraham, and H. H. Birnie. 1988. The necessity of each phase of the learning cycle in teaching high school physics. *Journal of Research in Science Teaching* 25: 39–58.

Renner, J. W., and W. C. Paske. 1977. Comparing two forms of instruction in college physics. *American Journal of Physics* 45 (9): 851–859.

Roth, K., and H. Garnier. 2006. What science teaching looks like: An international perspective. *Educational Leadership* 64 (4): 16–23.

Roth, K. J., H. Garnier, C. Chen, M. Lemmens, K. Schwille, and N. I. Z. Wickler. 2011. Videobased lesson analysis: Effective science PD for teacher and student learning. *Journal of Research in Science Teaching* 48 (2): 117–148.

Saunders, W. L., and D. Shepardson. 1987. A comparison of concrete and formal science instruction upon science achievement and reasoning ability of sixth-grade students. *Journal of Research in Science Teaching* 24 (1): 39–51n.

Schneider, L. S., and J. W. Renner. 1980. Concrete and formal teaching. *Journal of Research in Science Teaching* 17 (6): 503–517.

# CHAPTER 2

# Reframing Instruction With Universal Design for Learning

*Delinda van Garderen, Cathy Newman Thomas, and Kate M. Sadler*

If you are a teacher or an administrator, it would not be surprising to find that the learners you interact with are diverse in many ways. Based on national demographic statistics for public schools (McFarland et al. 2017), you most likely have, or are going to have, learners who have a disability (13%), such as children who are diagnosed with learning disabilities, autism, emotional and behavioral disorders, or intellectual disabilities; students who are English language learners (9.4%); or learners who come from culturally and linguistically diverse backgrounds (50%). Furthermore, you have, or are going to have, students who vary tremendously in their ability to learn, their interests, their skills, their preferences, and so on.

Maybe you have heard or read about the call in the *Next Generation Science Standards* (*NGSS*) that science is to be for "all students" (NGSS Lead States 2013, Appendix D). But who exactly are "all students"? According to the Every Student Succeeds Act (2015), *all students* represents a diverse group of learners, including those who are culturally and linguistically diverse, have a disability, and/or have limited English proficiency. Furthermore, it includes those who are considered gifted and talented. It is important to recognize that although there are many categories of diversity, learners may actually belong to multiple groups of diversity, and each group is a mixed group of learners with a variety of differences (NGSS Lead States 2013, Appendix D).

> **STOP AND CONSIDER ...**
>
> Before you read the rest of this chapter, take a moment to think about the students you are currently working with or may be working with in the future. Write down a list of ideas about them in response to the following questions. Who are your diverse learners? How are they alike? How are they different? What strengths do they bring to science? What challenges do they experience when learning science?
>
> As you think about your students, answer the following questions: Do you believe all your students can engage in science and learn challenging science concepts? Do your expectations differ for different students or groups of students? Do you feel comfortable teaching science to all of your students?

## Science for All?

According to the *NGSS*, not only is science to be for "all students," but there is also the expectation that teachers are to create learning opportunities that enable all of their students to meet "all standards" (NGSS Lead States 2013, Appendix D). Historically, expectations have varied for different groups of learners' participation in science, and not all learning opportunities have been equitable for all learners (McGinnis 2003). Here's the thing ... there is a lot of research (e.g., Aydeniz et al. 2012; Brusca-Vega, Brown, and Yasutake 2011; Therrien et al. 2011; 2014) that suggests that—given equitable learning opportunities—diverse learners are actually quite "capable of engaging in scientific practices and meaning-making in both science classrooms and informal settings" (NGSS Lead States 2013, Appendix D).

> **STOP AND CONSIDER ...**
>
> Can instruction be designed in such a way as to promote access to high-level, inquiry-based science learning for *all* learners, regardless or even in spite of their diverse needs?

We would not be surprised if many of you find it challenging to answer the above question. We suspect you are reading this book because you may be wondering how to teach science to such diverse learners! Guess what? We have wondered about this too—a lot. Therefore, the purpose of this chapter is to help you think about your instruction for diverse learners and how to use Universal Design for Learning (UDL) to (re)frame your instruction. This chapter outlines the framework of UDL, which can provide you with a systematic, intentional, and structured way to improve your instruction. With UDL, you can be more effective and impactful in how you plan for, include, and support diverse learners in doing meaningful science in every lesson and activity.

## What Is Universal Design for Learning?

The concept of UDL came from ideas originally developed by architects who recognized that many buildings and equipment were not accessible or easy to use for all individuals; in other words, there were many barriers for users (McGuire, Scott, and Shaw 2006). As a result, architects advocated that *all* users should be considered, proactively, in the planning and design stages of buildings, public use areas, equipment, and technology designed for public use. The designers embedded "solutions" into buildings and equipment to enable full access and use for everyone. Ronald Mace is credited with developing the concept that came to be called "universal design" (McGuire, Scott, and Shaw 2006).

Interestingly, when users with varied and diverse needs were considered during planning, it was found that these solutions were also beneficial for many others beyond those initially identified. For example, accessibility standards, such as a curb cut or ramps, may benefit a worker carrying boxes as well as a person in a wheelchair. People interested in the learning sciences borrowed the concept of universal design and worked to adapt it for the field of education. So, too, UDL promotes the idea of considering the needs of a wide range of learners *as you plan and design* materials and instruction (Gronneberg and Johnston 2015). Guiding this process is the UDL framework (Rose and Meyer 2002).

Before moving forward, it is important to acknowledge that educational researchers have been developing ideas about UDL for more than 20 years. The particular UDL framework used in this book is grounded in work developed by researchers at the Center for Applied Special Technology (CAST; 2018). There are numerous updated resources and materials that expand on ideas that we share; they can be found on CAST's website (*www.CAST.org*) or on the website of the National Center on Universal Design for Learning (*www.udlcenter.org/ aboutudlcenter*). We strongly encourage you to check them out.

## The UDL Framework: Principles, Guidelines, and Checkpoints

At the heart of the UDL framework are three principles, which draw from broad research findings in fields such as neuroscience and cognitive science (CAST 2018). Essentially, there are areas in our brain that are connected to *recognition* networks, which guide the "what" of learning; *strategic* networks, which guide the "how" of learning; and *affective* networks, which guide the "why" of learning. What the research suggests is that our instruction needs to address all three networks (CAST 2018; Rose and Strangman 2007). In particular, we need to focus on helping learners gather ideas and organize those ideas in a meaningful way, as well as provide ways for the learners to express their ideas, in a manner that engages them in the topic and supports them to stay motivated and engaged throughout the learning experience (CAST 2018; Rose and Strangman 2007).

Within each principle are three guidelines (see Figure 2.1), which have two key functions. First, each guideline highlights where potential barriers may exist that can prevent learners from accessing learning. Second, each provides concrete, evidence-based solutions (the checkpoints) to reduce or remove barriers that may be preventing diverse learners from accessing instruction and learning. Remem-

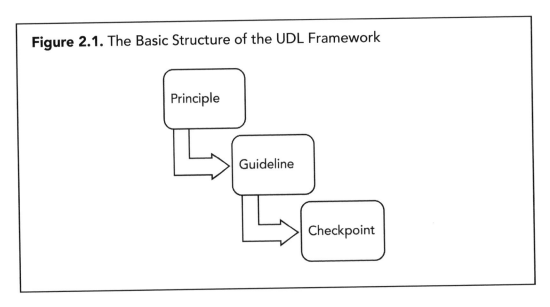

**Figure 2.1.** The Basic Structure of the UDL Framework

ber, *NGSS* calls for "all students, all standards," and for students with disabilities, the Individuals with Disabilities Education Act (2004)—and, more recently, the Every Student Succeeds Act (2015)—guarantees access to the general curriculum.

In the following sections, we describe the principles, guidelines, and checkpoints and provide examples to demonstrate how the principles can guide the development and implementation of inquiry-based science curricula. We start with the principle of providing multiple means of *representation* (Principle I), followed by the principle of multiple means of *action and expression* (Principle II), and lastly move on to multiple means of *engagement* (Principle III). Table 2.1 provides an overview of the UDL guidelines; however, the complete and expandable version of the UDL guidelines, along with many additional resources, can be found on the National Center for UDL website (*http://udlguidelines.cast.org*).

| **TABLE 2.1. Universal Design for Learning: Principles, Guidelines, and Checkpoints** | | |
| --- | --- | --- |
| **Provide Multiple Means of _Engagement_** | **Provide Multiple Means of _Representation_** | **Provide Multiple Means of _Action and Expression_** |
| Provide options for **recruiting interest** (7) <br>• Optimize individual choice and autonomy (7.1) <br>• Optimize relevance, value, and authenticity (7.2) <br>• Minimize threats and distractions (7.3) | Provide options for **perception** (1) <br>• Offer ways of customizing the display of information (1.1) <br>• Offer alternatives for auditory information (1.2) <br>• Offer alternatives for visual information (1.3) | Provide options for **physical action** (4) <br>• Vary the methods for response and navigation (4.1) <br>• Optimize access to tools and assistive technologies (4.2) |
| Provide options for **sustaining effort and persistence** (8) <br>• Heighten salience of goals and objectives (8.1) <br>• Vary demands and resources to optimize challenge (8.2) <br>• Foster collaboration and community (8.3) <br>• Increase mastery-oriented feedback (8.4) | Provide options for **language and symbols** (2) <br>• Clarify vocabulary and symbols (2.1) <br>• Clarify syntax and structure (2.2) <br>• Support decoding of text, mathematical notation, and symbols (2.3) <br>• Promote understanding across languages (2.4) <br>• Illustrate through multiple media (2.5) | Provide options for **expression and communication** (5) <br>• Use multiple media for communication (5.1) <br>• Use multiple tools for construction and composition (5.2) <br>• Build fluencies with graduated levels of support for practice and performance (5.3) |
| Provide options for **self-regulation** (9) <br>• Promote expectations and beliefs that optimize motivation (9.1) <br>• Facilitate personal coping skills and strategies (9.2) <br>• Develop self-assessment and reflection (9.3) | Provide options for **comprehension** (3) <br>• Activate or supply background knowledge (3.1) <br>• Highlight patterns, critical features, big ideas, and relationships (3.2) <br>• Guide information processing and visualization (3.3) <br>• Maximize transfer and generalization (3.4) | Provide options for **executive functions** (6) <br>• Guide appropriate goal-setting (6.1) <br>• Support planning and strategy development (6.2) <br>• Facilitate managing information and resources (6.3) <br>• Enhance capacity for monitoring progress (6.4) |
| _Goal: Expert learners who are ..._ | | |
| **Purposeful and motivated** | **Resourceful and knowledgeable** | **Strategic and goal directed** |

_Source:_ Modified from CAST (2018), Universal Design for Learning Guidelines, Version 2.2, graphic organizer found at _http://udlguidelines.cast.org._

You may notice in the table that multiple means of engagement (Principle III) is presented first, but in other documents it is presented third. The change in order was carried out as a way to highlight the essential role that engagement plays in learning—that, if a learner is not engaged, he or she is less likely to attend to the information being presented (CAST 2018). Like CAST, we consider all three principles to be of equal importance and believe that they should be considered simultaneously when applying the principles of UDL in your instruction. More importantly, when we reference numbers connected to the guidelines and checkpoints throughout this book, we aim to correspond them with what is presented in Table 2.1.

## Principle I: Representation

Central to any subject area, including science, is the content, which we can refer to as the "what" of learning. The what of learning in science connects to concepts and practices that students should know and develop that contribute to becoming resourceful and knowledgeable science learners (see Figure 2.2; CAST 2018). In our classrooms, the content of science may be presented (or *represented*) to learners in numerous ways such as printed materials, videos, models, lectures, discussions, and so on. However, how learners perceive and comprehend the information that is presented to them in these various formats may differ (CAST 2018). Not all formats are optimal, or even accessible, for all learners. For example, a learner who is hard of hearing may be unable to hear a group discussion well enough to support comprehension, or a learner who has a learning disability in reading may have difficulty reading the printed materials about the content. As a result, it may be necessary to provide options for representation that meet individual student needs (CAST 2018).

---

**Figure 2.2.** Connections Between UDL Principle I and the *NGSS*

In science, the "what" of learning connects to the *NGSS* in the following ways:

- **Disciplinary core ideas**—for example, PS3.A, "Definitions of Energy": The faster a given object is moving, the more energy it possesses (4-PS3-1).
- **Science and engineering practices**—for example, "Asking Questions and Defining Problems": Ask questions that can be investigated and predict reasonable outcomes based on patterns such as cause-and-effect relationships.
- **Crosscutting concepts**—for example, "Cause and Effect": Cause-and-effect relationships are routinely identified and used to explain change (4-ESS3-1).
- **Nature of science connections**—for example, "Science Is a Human Endeavor": Science affects everyday life (4-PS3-4).

---

## STOP AND CONSIDER ...

Recall the list you made earlier of your diverse learners. In what way might the content or instructional practices you are using be a barrier for some or all of the diverse learners in learning concepts and skills (the "what") during your science instruction? Below is a fillable chart to help you identify barriers. We have provided a couple of examples to help you think about your diverse learners.

### CHART FOR IDENTIFYING BARRIERS

| Student | Learning Needs and Strengths | Barrier for Accessing Content |
|---|---|---|
| Jada | Learning disability—2 years below grade level in reading but can understand text | Reading science textbook |
| Richard | Hearing impairment—needs amplification, enjoys working with others | Hearing lecture or listening to videos |
|  |  |  |
|  |  |  |
|  |  |  |
|  |  |  |
|  |  |  |

We present three guidelines and the corresponding checkpoints for providing multiple means of representation (see *http://udlguidelines.cast.org/representation*). The guidelines are (1) options for perception; (2) options for language, mathematical expression, and symbols; and (3) options for comprehension.

### Guideline 1: Perception

Content in science is presented in many different ways, requiring the use of our senses—sight, hearing, smell, touch, and taste. It is not possible to learn the content if it is imperceptible to the learner. As a result, we may need to use various representational forms to present the content.

Three checkpoints within this guideline help us think of different ways to present the same science information to make it accessible for all learners:

**1.1** **Offer ways of customizing the display of information.** For example, enlarging the text, using a different font, slowing down the rate of speaking during a lecture, or using pictures of video may increase accessibility, based on student need.

**1.2** **Offer alternatives for auditory information.** For example, providing diagrams, pictures, physical models, or other visual supports, or enabling speech-to-text and captioning technology features, may be more accessible for some students.

**1.3** **Offer alternatives for visual information.** For example, providing text-to-speech technology—including spoken descriptions of images, video, and physical models—and enabling haptic experiences (where students are able to touch, manipulate, and even build) with models can improve access to science content.

### Guideline 2: Language and Symbols

When you are providing content in science, the content has to be "decoded" or "translated" to be comprehended. Learners vary in their ability to access the content and may need additional support to make sense of text, numbers, symbols, and language.

We present five checkpoints within this guideline to offer solutions for helping learners access the content:

**2.1** **Clarify vocabulary and symbols.** For example, embed hyperlinks, footnotes, illustrations, diagrams, definitions, and explanations to help students with vocabulary words and symbols in a text.

**2.2** **Clarify syntax and structure.** For example, highlight transition words in a text to make relationships explicit in the text.

**2.3** **Support decoding of text, mathematical notation, and symbols.** For example, use text-to-speech and peer partners or a small-group read-aloud.

**2.4** **Promote understanding across languages.** For example, provide key vocabulary and pronunciations in the dominant or heritage language.

**2.5** **Illustrate through multiple media.** For example, provide key ideas in one representational form simultaneously paired with another representational form.

### Guideline 3: Comprehension

A knowledgeable learner in science is one who understands or makes sense of what has been presented and then can use what he or she has learned and apply it in meaningful ways within the context. A knowledgeable learner can also transfer the new learning to other contexts where it is useful. Students vary in their ability to take information that is presented to them and transform it into usable knowledge (Bransford, Brown, and Cocking 2000; Rose and Gravel 2009).

We present four checkpoints within this guideline that can be used to help learners comprehend the content:

3.1     **Activate or supply background knowledge.** For example, use advanced organizers such as a concept map.

3.2     **Highlight patterns, critical features, big ideas, and relationships.** For example, call attention to important features when observing, use multiple examples and non-examples as a way to draw out critical features, or provide a graphic organizer that models relationships.

3.3     **Guide information processing and visualization.** For example, chunk information around key ideas and categories, or release information in smaller pieces to facilitate processing and organization of ideas.

3.4     **Maximize transfer and generalization.** For example, provide a guided learning experience that supports the learner in generalizing a key concept to a new situation.

In the Classroom Snapshot that follows—"What's the 'Matter,' Anyway?"—a fifth-grade teacher describes an experience with the *representation* principle.

## CLASSROOM SNAPSHOT

### What's the "Matter," Anyway?
Cody Sanders, grade 5 teacher

***5-PS1-1.*** *Develop a model to describe that matter is made of particles too small to be seen.*

***5-PS1-4.*** *Conduct an investigation to determine whether the mixing of two or more substances results in new substances.*

As a preservice teacher at the University of Missouri–Columbia, I often heard variations of the phrase, "Most times you could plan a lesson exactly the way you want to, but it could go haywire once put into place." At the time, I never really thought about it. I was just excited to use all of the knowledge and skills that I had learned from a science methods course in which I was introduced to the 5E Learning Cycle, a special education course in which I had learned about UDL, and my experience working alongside veteran teachers in the Quality Elementary Science Teaching program. I thought I was ready to take on the world. I felt like I could genuinely create lessons with all of my learners in mind. I thought I was cognizant of what I needed to do and why I needed to do it.

Not until teaching about solubility on my own as a student teacher did I truly experience how a lesson can change out of the blue! I quickly learned the importance of planning a lesson that builds or creates a storyline that makes sense while keeping the success of all learners in mind.

Fast-forward two years, and now I am a fifth-grade teacher with my own classroom. I am teaching the same lesson again. Fortunately, I get to take another swing and to improve upon aspects of the lesson that were not at the forefront when I initially planned and taught it as a student teacher. In particular, I want to plan for my students' needs and the need for a conceptual storyline in the content I am going to teach. I focus on strengthening the conceptual storyline in three phases of the 5E Learning Cycle (Explain, Extend, and Evaluate) and embedding UDL solutions targeted to the learning needs of my students.

One principle that I particularly focus on in my planning is *representation*. In my original lesson, I noticed two problems: (1) Some of my students did not have the necessary background information and understanding to access the material (Guideline 3, Checkpoint 3.1); and (2) some students were distracted (i.e., by the color of the salt or sugar) from important aspects of the phenomenon (solubility), which prevented them from fully understanding the task, and they needed to

*(continued)*

*(continued)*

have the phenomenon represented again before moving on (Guideline 3, Checkpoint 3.2). To address these barriers, I illustrate the phenomenon multiple times and supply background knowledge via videos (see the box below).

---

**CODY'S UDL ACTION STEPS**

Principle: Representation

Guideline: Language and symbols (barrier)
- Checkpoint 3.1: Activate or supply background knowledge

Guideline: Comprehension (barrier)
- Checkpoint 3.2: Highlight patterns, critical features, big ideas, and relationships

---

The table below shows how I originally planned two phases of the 5E Learning Cycle and then the changes I made when planning to reteach this learning cycle with my fifth-grade students.

| CHANGES TO CODY'S LESSON PLANNING | | |
|---|---|---|
| **Learning Cycle Phase** | **Original Lesson** | **Two Years Later** |
| Explain | Students pinch one color of salt/sugar into hot water and a different color of salt/sugar into cold water. They observe and record what is happening and they create a model. | After completing the experiment, we also watched a video of what we had just witnessed. (I added the video because students in the original lesson fixated on the color of the salt/sugar instead of on the concept of solubility in warm vs. cold water. As they watched, I explicitly pointed out the differences in temperatures and the differing dissolving rates of the sugar/salt.) |
| Extend | Students work through a real-life scenario on why it is better to add sweetener to a beverage before putting ice into the drink. | Before completing the real-life scenario, we discussed how sweet tea is brewed and watched a video. (I added this step because not all students in the original lesson knew how tea was brewed and they had difficulty accessing the material, which distracted them from focusing on the concept of solubility in warm vs. cold water.) |

As a student teacher, planning was one of the toughest parts, and not everything may go to plan. But I did learn that the 5E Learning Cycle and the UDL principles are critical when teaching science, or any subject for that "matter." This combined model may take time to use when planning; however, it is a perfect roadmap that can result in meaningful learning for all of your students.

## Principle II: Action and Expression

The ability to demonstrate what has been learned in any subject area, including science, is the primary way we gauge learner knowledge. Within this principle, the focus is on ensuring that learners can fully communicate what they know through some action, which we can evaluate to ensure that objectives are met. This is also referred to as the "how" of learning. The purpose of this principle is to provide opportunities for students to develop strategies (e.g., self-reflection, planning, goal setting) and skills (e.g., speaking, writing, assembling a physical model) that enable them to demonstrate what they know.

Learners in our classrooms vary in skills and strategies and the ability to navigate the learning environment and express what they know (Rose and Gravel 2009). For example, some students may struggle to hold and manage writing with a pencil; some may struggle to organize information, even information that they comprehend well; and some may not be able to fully communicate their thinking orally. However, it is possible to create a learning environment that is navigable and interactive for all learners, and you can provide supports that enable students to demonstrate what they know when given alternatives for actions and expression.

### STOP AND CONSIDER ...

Recall the list you made earlier of your diverse learners. In what way might the content or instructional practices you are using be a barrier for some or all of the diverse learners in demonstrating what they have learned and know (the "how")? Below is a fillable chart to help you identify barriers. We have provided a couple of examples to help you think about your diverse learners.

| CHART FOR IDENTIFYING BARRIERS | | |
|---|---|---|
| **Student** | **Learning Needs and Strengths** | **Barrier for Demonstrating What They Know** |
| Juan | Learning disability—needs help with organizing ideas to write, but very creative | Sequencing ideas to write an essay or report |
| Asha | Cerebral palsy—poor fine-motor skills, strong reading skills | Using pen or pencil to write or draw diagrams and models |
| | | |
| | | |
| | | |
| | | |

We present three guidelines and the corresponding checkpoints for providing multiple means of action and expression (see *http://udlguidelines.cast.org/action-expression*). The guidelines are (1) options for physical action, (2) options for expression and communication, and (3) options for executive function.

### Guideline 4: Physical Action

The materials and equipment that we provide our learners to use in science often permit limited forms of navigation and interaction. For example, when they read a book, it requires that they turn a page. Or, when writing a story, they may only have a pencil or keyboard to use.

There are two checkpoints that can be used to identify different ways for learners to physically respond:

4.1　**Vary the methods for response and navigation.** For example, offer choices or alternatives for physical interaction with materials such as a keyboard, adapted keyboard, voice, switches, joystick, manipulatives, and physical objects.

4.2　**Optimize access to tools and assistive technologies.** For example, consider alternative keyboards, switch and scanning options, and overlays for touch screens.

### Guideline 5: Expression and Communication

There are numerous forms of communication available. However, not all forms of communication can be used by all learners. For example, a student with challenges in expressive language—who might have word-finding difficulties, a stutter, or selective mutism—could not give, or might not be comfortable giving, an oral report. A learner who has cerebral palsy affecting motor movement may not be able to use a pen to write a report. And a learner with a learning disability in writing may struggle to write an essay or complete a lab report but could communicate very well orally. It should not be assumed that because a student cannot respond in one format, he or she does not understand the content. Therefore, it is important where necessary to provide alternative ways for learners to communicate and express their ideas—including during assessment.

Three checkpoints highlight the different ways to provide options for enabling learners to express and communicate their ideas:

5.1　**Use multiple media for communication.** For example, use speech, text, dance, drawing, video, music, or visual art.

5.2　**Use multiple tools for construction and composition.** For example, consider sentence starters, sentence strips, text-to-speech or speech-to-text software, virtual or concrete manipulatives, spell-checkers, or web applications.

**5.3** **Build fluencies with graduated levels of support for practice and performance.** For example, provide a model and use it to add additional and more complex features one-by-one to scaffold students into higher levels of thinking; use different models or examples; or embed scaffolds that can be gradually released as the learner becomes increasingly independent in a task.

### Guideline 6: Executive Functions

A learner's ability to carry out a task in a skillful manner is dependent on *executive functions*, which are the processes we use to guide and direct ourselves as we carry out tasks (Center on the Developing Child 2011). One way to look at executive function is to think of it as the "air traffic control system" of our brain (Center on the Developing Child 2011). Executive function encompasses functions for goal setting, planning, monitoring, organization, memorization, and strategy use. These functions are necessary to take advantage of the learning opportunities that are presented and provided. Many learners experience difficulties with executive functions. For example, students with learning disabilities often have difficulty knowing what strategy to use to solve a problem or have limited knowledge of strategies for carrying out a task such as writing a report. Students with attentional issues often have difficulty inhibiting their focus to the task at hand and get easily distracted with irrelevant information, resulting in a lack of focus in starting, organizing, self-monitoring, and completing science assignments. Therefore, the purpose of this guideline is to provide supports for learners that help them carry out executive tasks themselves (e.g., how to plan, how to organize, how to monitor, and how to evaluate).

Four checkpoints highlight the various ways to provide supports to enhance executive functioning:

**6.1** **Guide appropriate goal-setting.** For example, post goals and objectives, and provide models or examples of goals.

**6.2** **Support planning and strategy development.** For example, teach and promote the use of strategies to help learners become more "plan-ful" for carrying out a task, such as project-planning templates, checklists, and guides. Peer coaches and mentors can model and support planning and problem solving.

**6.3** **Facilitate managing information and resources.** For example, encourage the use of graphic organizers, templates, checklists, guides for note-taking, and categorizing systems.

6.4 **Enhance capacity for monitoring progress.** For example, use models of self-assessment strategies, self-question prompts for self-monitoring and feedback, and assessment checklists.

## UDL in Action

The Classroom Snapshot below, "An Energetic Discussion," demonstrates Principle II in action. As you read the snapshot, think about whether the teacher's solution would work for your diverse learners. How might you implement her ideas in your instruction? What additional solutions could you build in?

## CLASSROOM SNAPSHOT

### An Energetic Discussion
Linda Buchanan, grade 4 teacher

***4-PS3-2.*** *Make observations to provide evidence that energy can be transferred from place to place by sound, light, heat, and electric currents.*

I am a firm believer that all children can learn when they are taught in a way that matches how they learn. To me, the 5E Learning Cycle does that. It breaks down the lesson into parts—and it builds from what students know to new ideas and understandings about science. To develop these ideas, students have to be able to share their ideas, consider the ideas of others, and revise their thinking. This is not something that children enter my class knowing how to do naturally.

In my classroom, I anticipated that a barrier for students would be participating in discussions as they worked in groups, particularly in the Explain phase of the lesson. Previous discussions had quickly deteriorated into accusations of who was "wrong" versus using evidence to support and evaluate ideas. The result was that some students got defensive, while others shut down.

To address this barrier (Guideline 5, Checkpoint 5.2), along with reminding the students about group norms, I provided sentence starters on strips of paper to use when sharing ideas in a group discussion (see the box, p. 36). Using this strategy helped me build a safe, nonthreatening learning community that encourages students to take risks. In addition, it provided a framework for building the real-life skill of communicating thoughts and ideas to others.

*(continued)*

(continued)

---

**LINDA'S UDL ACTION STEPS**

Principle: Action and expression

Guideline: Expression and communication (barrier)
- Checkpoint 5.2: Use multiple tools for construction and composition (solution)

Action step: Sentence starters for discussion—
1. I agree with you because …
2. I disagree with you because …
3. I understand why you think that …
4. This makes sense, I was thinking the same thing also …
5. So, I think I hear you saying …

---

During a lesson in which students were constructing electrical circuits, I listened to one group of students discuss the role of the battery in the circuit:

Dante:    *I think batteries are used to give off the energy needed to light something up.*

Kim:    *I disagree; batteries do not give off energy. Energy is stored in a battery until you need it.*

Malcolm:    *I think I understand why you said that, because if it constantly gave off energy, there would be nothing left to power the light.*

In another group, a similar conversation was taking place:

Kanye:    *I disagree. If the battery is the energy source, why is it not hot when we touch it?*

Shaun:    *It is not hot because it is not doing any work. Touch the bulb, Bro.*

Using the sentence starters helped the students have deeper discussions of their ideas and the reasoning behind them. In these cases, students were beginning to reason about evidence of energy transfer.

With practice, the students began to apply these sentence starters for discussion during "teachable moments" shared with the whole class. For example, Shaun went to the battery cubby and picked up two extra batteries, then connected them to the bulb. "I made the bulb shine brighter," he announced. I recognized this as a perfect opportunity to extend the students' thinking and connect back to the idea that the battery is the source of energy in the circuit.

I asked Shaun to explain to the class what he did and why he decided to try more batteries. I then invited the class to respond to his ideas.

*(continued)*

*(continued)*

Shaun: *I thought if you had more batteries there would be more energy, so I connected them together like this [he demonstrates].*

Teacher: *What do the rest of you think about that?*

Jimmy: *I think I heard you say that the more batteries you have, the more energy you have. I see, it's like you add the two batteries together so there is more energy.*

Teacher: *Do you think having two batteries would always make the bulb brighter?*

Lein: *Our group had a battery that wasn't doing anything. It didn't get warm and the bulb didn't light.*

Shaun: *Maybe it's a dead battery—it doesn't have any more energy in it.*

Teacher: *How can you tell if a battery is dead?*

Lein: *If you connect all the wires in the right way and it doesn't light, it could be a dead battery.*

It took a lot of practice, but eventually I observed that students facilitated each other's learning by promoting positive thinking and using positive responding. As one of my students commented, "Before, I would just speak without thinking about what I said. Now I have to stop and think and say something positive about the other person's ideas and use evidence to back up what I say."

## Principle III: Engagement

Central to learning is learner affect (CAST 2018; Pintrich 2003; Trujillo and Tanner 2014; Weber 2008). A learner's affect often impacts and influences how engaged and motivated he or she is to learn, including in science. Our goal as teachers is to help our students become purposeful, motivated learners (CAST 2018), eager to engage with, participate in, and learn about science. The purpose of this principle, then, is to help students not only become interested in what they are learning, but also sustain their motivation to learn and regulate themselves as learners. Creating safe and welcoming science learning environments that support students' coping skills, allow them to explore and test their ideas, and are motivating and engaging—this can be thought of as the "why" of learning.

Students vary tremendously in their response to the learning environment. Some students, for example, prefer to work with others, while others like to work alone or even have real difficulty working with their peers. Not all learners are engaged by the same activities, and while some learners enjoy more open-ended or problem-based learning opportunities, others do not, preferring more structured activities and a predictable routine (CAST 2018). Additionally, some students, and especially those with disabilities, may not be confident in their thinking, and they may be unwilling to take the risk of making an error publicly in a classroom (Lackaye and Margalit 2006; Morrison and Cosden 1997). This prevents them from fully engaging in science, which often focuses on testing and discarding ideas. As challenging as this might be, to the extent possible, we need to create a learning environment that engages all of our learners and creates and instills a desire to learn about science.

## STOP AND CONSIDER …

Recall the list you made earlier of your diverse learners. In what way might the content or instructional practices you are using be a barrier for some or all of the diverse learners in engaging, sustaining, and regulating their learning? Below is a fillable chart to help you identify barriers. We have provided a couple of examples to help you think about your diverse learners.

| CHART FOR IDENTIFYING BARRIERS | | |
|---|---|---|
| Student | Learning Needs and Strengths | Barrier for Engaging, Sustaining, and Regulating Learning |
| Jaden | Gifted—loves science and engineering | Content too easy |
| Craig | ADHD—inattentive, enjoys constructing models and drawing | Sustaining attention during tasks and missing critical concepts |
| | | |
| | | |
| | | |
| | | |
| | | |

We present three guidelines and the corresponding checkpoints for providing multiple means of action and expression (see *http://udlguidelines.cast.org/engagement*). The guidelines are (1) options for recruiting interest, (2) options for sustaining effort and persistence, and (3) options for self-regulation.

### Guideline 7: Recruiting Interest

There are many ideas in science that we expect our learners to know and understand. However, if students are not attending to or engaging in the content, they will not be learning (CAST 2018). As teachers, we are responsible for capturing every student's interest and attention. This can be challenging given that we all differ in what we find interesting. Further, our interests tend to change over time.

Three checkpoints highlight the various ways we might go about recruiting students' interest for learning, including science learning:

**7.1** **Optimize individual choice and autonomy.** For example, provide choices in the tasks such as levels of challenge, context for the task, content, tools to gather or produce information, timing and sequencing, and ways to present information.

**7.2** **Optimize relevance, value, and authenticity.** For example, make lessons culturally relevant, socially relevant, and personalized to the learners' lives.

**7.3** **Minimize threats and distractions.** For example, vary the level of novelty or risk via the use of such things as a schedule, cues, and alerts; monitor the level of social demand; and minimize the level of sensory stimulation such as amount of noise.

### Guideline 8: Sustaining Effort and Persistence

A novel activity may initially capture students' attention and engage them as learners. However, many kinds of learning require sustained attention and effort (CAST 2018). Tasks that are well chosen or developed can initially engage learners; however, students do differ considerably in their persistence for learning and their ability to sustain attention and effort, for a variety of reasons. Further, not all students have the required levels of self-regulation and self-determination to persist. There are many strategies available to help teachers support learners in developing and improving their self-regulation and self-determination skills, enabling them to increase their focus and remain on task.

Four checkpoints provide direction for helping students persist in their interest and efforts and remain engaged as learners:

**8.1** **Heighten salience of goals and objectives.** For example, display the goal in different ways, prompt learners to restate the goal, and break down goals into smaller or short-term objectives.

**8.2** **Vary demands and resources to optimize challenge.** For example, differentiate the degree of difficulty or complexity of the task and provide alternatives in the way to carry out a task.

8.3 **Foster collaboration and community.** For example, use cooperative learning groups and peer-tutors, and create norms and expectations for group work.

8.4 **Increase mastery-oriented feedback.** For example, provide frequent and timely feedback specific to such things as task progress and completion, effort, improvement, perseverance, positive strategies for future success, substance, and information.

### *Guideline 9: Self-Regulation*

While a good part of our focus should be on developing tasks that promote learning, it may be necessary to implement strategies that help some learners better self-regulate—"to strategically modulate one's emotional reactions or states in order to be more effective at coping and engaging with the environment" (CAST 2018). Many students develop self-regulatory behaviors and skills on their own; however, some learners have significant difficulties in managing this reflective and personal self-oversight. A critical part of teaching is implementing strategies that support students in learning to monitor and manage their own engagement and affect to enable them to more fully and consistently engage in the learning opportunities.

Three checkpoints provide good guidance as to where supports can be embedded within instruction:

9.1 **Promote expectations and beliefs that optimize motivation.** For example, provide mentors (e.g., scientists); encourage self-reflection and identification of personal goals (e.g., content to be learned); and provide supports such as checklists or guides that focus on on-task behavior, self-monitoring and management (e.g., completion of tasks), regulation of behavior, self-reflection, and self-reinforcement (e.g., "I listened to others in my group").

9.2 **Facilitate personal coping skills and strategies.** For example, provide models and scaffolds for managing frustration (e.g., break the task into steps), recruiting and receiving emotional support, developing coping skills (e.g., provide guidelines or examples for how to ask for help), and maintaining positive beliefs and self-efficacy.

9.3 **Develop self-assessment and reflection.** For example, chart and display data of behavior (e.g., time on task, number of tasks completed) in order to monitor these behaviors.

## UDL in Action

The Classroom Snapshot below, "Rocket Science," demonstrates Principle III in action. As you read the snapshot, think about whether the teacher's solution would work for your diverse learners. How might you implement his ideas in your instruction? What additional solutions could you build in?

## CLASSROOM SNAPSHOT

### Rocket Science

Warren Soper, grade 4 teacher

***4.ETS1.A.1.*** *Define a simple design problem reflecting a need or a want that includes specified criteria for success and constraints on materials, time, or cost.*

***4.ETS1.B.1.*** *Generate and compare multiple possible solutions to a problem based on how well each is likely to meet the criteria and constraints of the problem.*

***4.ETS1.C.1.*** *Plan and carry out fair tests in which variables are controlled and failure points are considered to identify aspects of a model or prototype that can be improved.*

I try to present science and engineering as a learning process, where ideas and designs are proposed, tested, and refined continually. Many students, however, are often used to "one-shot" opportunities, in which the stakes are succeed or fail. Kids think they have to do everything right the first time, and if they don't achieve what they think others expect from them, then they are not doing as well as they should—and they fail. (Unfortunately, many adults feel this way too!) To me, we learn through failure—by analyzing what doesn't work as we expect or intend. Often, students see failure as an end to an activity, not as a stepping-stone to the next phase of the project.

Some of my students have particular difficulty coping with an unsuccessful attempt in any endeavor and will give up rather than continuing to try to reach a successful completion of the task.

Building on what students learn about motion in third grade, our fourth-grade students design and test rockets in an engineering design project. To help the students understand the importance—and benefit—of not accepting failure, I start the unit by showing videos of very early unsuccessful rocket launches (Guideline 9, Checkpoint 9.1). The students are amazed to see rockets fall over, fail to launch, or explode in mid-flight (see the box, p. 42).

*(continued)*

*(continued)*

**WARREN'S UDL ACTION STEPS**

Principle: Engagement

Guideline: Self-regulation (barrier)
- Checkpoint 9.1: Promote expectations and beliefs that optimize motivation (solution)
- Checkpoint 9.2: Facilitate personal coping skills and strategies (solution)

Action steps:
1. Mentors and self-reflection: Rocket launch video
2. Models: Scientists and engineers

I emphasize to the students that these early scientists and engineers whose designs failed were in fact some of the best, most innovative minds on the planet at that time. They did not just give up when they failed but instead asked themselves what they might do differently so that they would eventually succeed (Guideline 9, Checkpoint 9.2). Having my students know that these brilliant men and women failed to successfully launch their first attempts makes it easier for the students to accept when their own initial rocket designs do not achieve their desired results. We often refer back to the video as we reflect on the performance of their rockets. As a result, I find that students are more motivated to engage in the redesign phase of the engineering design process!

When we call something "rocket science," we often mean something that is too difficult to understand or achieve—in my classroom, students come to know rocket science as something they can do, as long as they don't give up!

## Next Steps

In Part II of this book, we provide you with many examples of how UDL can be integrated within a well-designed 5E Learning Cycle. As you read the lesson vignettes, look for examples and ideas that you could apply in your classroom situation. Consider highlighting examples you might find beneficial for your diverse learners, using a color to match the principle (e.g., pink for representation, blue for action and expression, and green for engagement). Make sure that you also check out Part III of the book, as we provide you with further ideas to use and with guidelines to help you think about when, where, and how to apply UDL in your classroom setting.

## Safety Considerations

Science teaching necessarily involves working with different materials, and at times, this can pose safety hazards. Safety always needs to be the first concern in all our teaching. Teachers need to be sure that their rooms, playgrounds, and other learning spaces are appropriate for the activities being conducted. That means that personal protective equipment (PPE) such as sanitized safety glasses with side shields or safety goggles are available and used properly. When students are using potentially harmful supplies and equipment, PPE is to be worn as appropriate during all components of investigations (i.e., the setup, hands-on investigation, and cleanup). At a minimum, the eye protection PPE provided for students to use must meet the ANSI/ISEA Z87 standard. Remember to also review and comply with all safety policies and procedures that have been established by the place of employment. Teachers also must practice the proper disposal of materials.

The National Science Teaching Association maintains a good website (*www.nsta.org/safety*) that provides guidance for teachers at all levels. The site also has a safety acknowledgment form (which is sometimes called a "safety contract") (*http://static.nsta.org/pdfs/SafetyAcknowledgmentForm-ElementarySchool.pdf*) specifically for elementary students to review with their teachers and have signed by parents or guardians. It cannot be overstated that safety is the single most important part of any lesson. Safety notes have been included with each lesson in this book to highlight specific concerns that might be associated with particular lessons.

The safety precautions associated with each investigation are based, in part, on the use of the recommended materials and instructions, legal safety standards, and better professional safety practices. Selecting alternative materials or procedures for these investigations may jeopardize the level of safety and therefore is at the user's own risk. Remember that an investigation includes three parts: (1) the setup, which is what you do to prepare the materials for students to use; (2) the actual investigation, which involves students using the materials and equipment; and (3) the cleanup, which includes cleaning the materials and putting them away for later use. The safety procedures and PPE that we stipulate for each investigation apply to all three parts.

## References

Aydeniz, M., D. F. Cihak, S. C. Graham, and L. Retinger. 2012. Using inquiry-based instruction for teaching science to students with learning disabilities. *International Journal of Special Education* 27 (2): 189–206.

Bransford, J. D., A. L. Brown, and R. R. Cocking. 2000. *How people learn: Brain, mind, experience, and school* (Expanded ed.). Washington, DC: National Academies Press.

Brusca-Vega, R., K. Brown, and D. Yasutake. 2011. Science achievement of students in co-taught, inquiry-based classrooms. *Learning Disabilities: A Multidisciplinary Journal* 17 (1): 23–31.

Center for Applied Special Technology (CAST). 2018. Universal Design for Learning guidelines (Version 2.2). *http://udlguidelines.cast.org*.

Center on the Developing Child. 2011. Building the brain's "air traffic control" system: How early experiences shape the development of executive function. Working Paper No. 11, Harvard University. *www.developingchild.harvard.edu.*

Every Student Succeeds Act of 2015, Pub. L. No. 114-95 § 114 Stat. 1177 (2015–2016).

Gronneberg, J., and S. Johnston. 2015. 7 things you should know about Universal Design for Learning (Brief). Washington, DC: Educause Learning Initiative. *www.educause. edu/library/resources/7-things-you-should-know-about-universal-design-learning.*

Individuals with Disabilities Education Act, 20 U.S.C. § 1400 (2004).

Lackaye, T. D., and M. Margalit. 2006. Comparisons of achievement, effort, and self-perceptions among students with learning disabilities and their peers from different achievement groups. *Journal of Learning Disabilities* 39 (5): 432–446. *https://doi. org/10.1177/00222194060390050501.*

McFarland, J., B. Hussar, C. de Brey, T. Snyder, X. Wang, S. Wilkinson-Flicker, S. Gebrekristos, J. Zhang, A. Rathbun, A. Barmer, F. Bullock Mann, and S. Hinz. 2017. *The condition of education 2017.* NCES 2017-144, U.S. Department of Education. Washington, DC: National Center for Education Statistics. *https://nces. ed.gov/pubsearch/pubsinfo.asp?pubid=2017144.*

McGinnis, J. R. 2003. The morality of inclusive verses exclusive settings: Preparing teachers to teach students with mental disabilities in science. In *The role of moral reasoning on socioscientific issues and discourse in science education*, ed. D. Zeidler, 196–215. Boston: Kluwer Academic Publishers.

McGuire, J. M., S. S. Scott, and S. F. Shaw. 2006. Universal design and its applications in educational environments. *Remedial and Special Education* 27 (3): 166–175. *https://doi.org/10.1177/07419325060270030501.*

Morrison, G. M., and M. A. Cosden. 1997. Risk, resilience, and adjustment of individuals with learning disabilities. *Learning Disability Quarterly* 20 (1): 43–60. *https://doi. org/10.2307/1511092.*

NGSS Lead States. 2013. *Next Generation Science Standards: For states, by states.* Washington, DC: National Academies Press. *www.nextgenscience.org/next-generation-science-standards.*

Pintrich, P. R. 2003. A motivational science perspective on the role of student motivation in learning and teaching contexts. *Journal of Educational Psychology* 95 (4): 667. *https://doi.org/10.1037/0022-0663.95.4.667.*

Rose, D., and J. W. Gravel. 2009. Getting from here to there: UDL, global positioning systems, and lessons for improving education. In *A policy reader in Universal Design for Learning*, ed. D. T. Gordon, J. W. Gravel, and L. A. Schifter, 5–18. Cambridge, MA: Harvard Education Press.

Rose, D. H., and A. Meyer. 2002. *Teaching every student in the digital age: Universal Design for Learning.* Alexandria, VA: Association for Supervision and Curriculum Development. (ERIC Document Reproduction no. ED 466 086).

Rose, D. H., and N. Strangman. 2007. Universal Design for Learning: Meeting the challenge of individual learning differences through a neurocognitive perspective. *Universal Access in the Information Society* 5 (4): 381–391. *https://doi.org/10.1007/ s10209-006-0062-8.*

Therrien, W. J., J. C. Taylor, J. L. Hosp, E. R. Kaldenberg, and J. Gorsh. 2011. Science instruction for students with learning disabilities: A meta-analysis. *Learning Disabilities Research & Practice* 26 (4): 188–203. *https://doi.org/10.1111/j.1540-5826.2011.00340.x.*

Therrien, W. J., J. C. Taylor, S. Watt, and E. R. Kaldenberg. 2014. Science instruction for students with emotional and behavioral disorders. *Remedial and Special Education* 35 (1): 15–27. *https://doi.org/10.1177/0741932513503557.*

Trujillo, G., and K. D. Tanner. 2014. Considering the role of affect in learning: Monitoring students' self-efficacy, sense of belonging, and science identity. *CBE– Life Sciences Education* 13 (1): 6–15. *https://doi.org/10.1187/cbe.13-12-0241.*

Weber, K. 2008. The role of affect in learning real analysis: A case study. *Research in Mathematics Education* 10 (1): 71–85. *https://doi.org/10.1080/14794800801916598.*

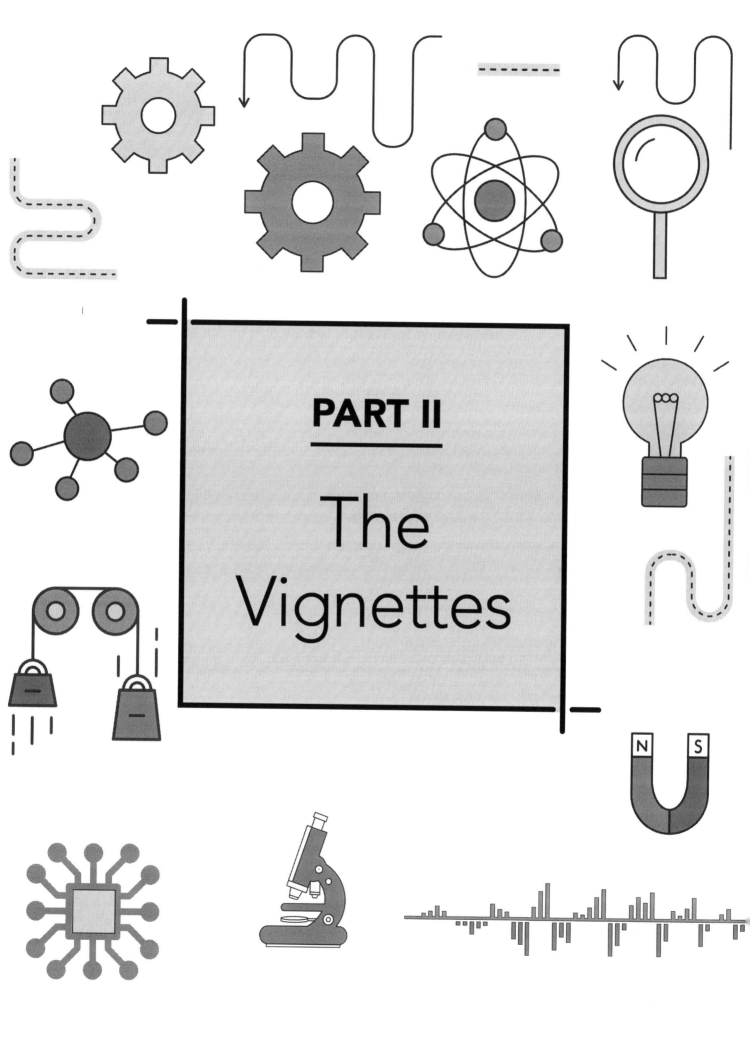

# PART II

## The Vignettes

# CHAPTER 3

# Why Does Matter *Matter*?

### *Shelli Thelen*

This chapter introduces a lesson that engages students in developing an initial understanding of matter by observing and comparing properties, noting patterns for different kinds of materials, and then using that knowledge to identify an unknown substance. This learning cycle represents the first lesson in a fifth-grade unit about matter that engages students in figuring out how a ship might have become stranded in a desert nearly 100 miles from water.[1] The emphasis is on properties that can be used to distinguish matter from non-matter and on differentiating solids, liquids, and gases. See Figure 3.1 (p. 50) for the conceptual storyline of this lesson, "What Is Matter?"

---

1. See *https://en.wikipedia.org/wiki/Moʻynoq*.

**Figure 3.1.** Conceptual Storyline of the Lesson

# What Is Matter?

**Engage**

Card-sort activity: Is it matter?

**Key idea:** "Matter" has different meanings in everyday life and science

**Linking question:** What do scientists consider "matter"?

**Explore**

Observing properties of materials

**Key idea:** Not everything is matter; matter has specific properties

**Linking question:** What properties distinguish matter from non-matter?

**Explain**

Analyzing patterns in our data

**Key idea:** All matter has mass and takes up space (volume)

**Evaluate**

Mysterious substance (oobleck)

**Key idea:** Some things have properties of more than one kind of matter

**Linking question:** Is all matter a solid, liquid, or gas?

**Extend**

Classifying matter by properties

**Key idea:** Matter can be solid, liquid, or gas; each category has unique properties

**Linking question:** What properties distinguish different types of matter from each other?

## LESSON VIGNETTE

After many years of teaching kindergarten, I was reassigned to fifth grade. In many ways, I was learning along with my students as I implemented new curricula, addressed new standards, and taught new topics. My first year of teaching fifth grade was a much-needed change from my many beloved years spent teaching kindergarten. I started the beginning of my first year in fifth grade teaching the science units on Earth systems and space systems with gusto, and I felt accomplished. Then November hit and the unit of matter stared at me square in the face. Matter. "What is matter?" It's a big question to think about. Matter has multiple meanings. Matter can be a noun, or it can be a verb.

Admittedly, I was reluctant to teach matter. I was lacking confidence in my own understanding of the content. Simply put, the progression and implementation of teaching matter did not feel cohesive. I gave myself grace that it would be impossible to be an all-star teacher in a new grade level in all content areas.

Talking with my grade-level team, I expressed my lack of enthusiasm for teaching matter and how I found it to be one of my greatest challenges in teaching fifth grade. I shared my urgency for support with planning the matter unit and my longing for a sense of cohesion in the progression of the unit. Mapping out the curriculum's progression is something that usually comes easily for me, and I was feeling frustrated. Fortunately, my team and I had the opportunity that summer to attend the Quality Elementary Science Teaching (QuEST) professional development program, where I learned about planning a coherent conceptual storyline.

Using a conceptual storyline is complementary to backward lesson plan design (Wiggins and McTighe 2005). It requires the teacher to think about the sequence and progression of lessons so that the conceptual knowledge builds with each experience across lessons, units, and topics. This intentional, thoughtful progression is often embedded in a meaningful context or real-life scenario through the 5E Learning Cycle. The matter unit that we developed during QuEST contains eight 5E learning cycles that engage students in developing their conceptual understanding about matter. The first learning cycle of the progression engages them to define matter and to recognize that matter can be identified by certain characteristics.

What I also learned in QuEST about the conceptual storyline was to attend to *the ideas that students would be developing during these activities.* I was used to planning by starting with the activities that the students would be participating in—for example, a card-sort, hands-on exploration, or data analysis. This storyline (shown in Figure 3.1) focuses on the essential question "What is matter?" by

*(continued)*

*(continued)*

starting from students' prior knowledge of this term in everyday life and helping them develop understanding of the use of the term in science. Students build this understanding through their own observations of properties and patterns in properties, with specific examples of matter and non-matter, as well as solids, liquids, and gases. While we traditionally teach these as hard and fast categories, the final activity helps students realize that sometimes matter can have properties of both solids and liquids.

Organizing the lesson in this way helped me focus on how students might develop their knowledge in a logical sequence, the questions they would need to answer to move forward in developing their ideas, and the activities that would help them construct their understanding. It also helped me consider how students would be using their knowledge, as described in the *Next Generation Science Standards* (*NGSS*) performance expectation (NGSS Lead States 2013), to identify matter based on its properties (see Table 3.1). That is, knowing the properties of matter *matter* because they are useful in identifying and classifying materials. Additionally, it helped me ensure that students had the necessary knowledge and skills to prepare them to explain phenomena related to matter in later lessons, such as melting, evaporating, and dissolving.

In the sections that follow, I outline the lesson as it unfolded in my classroom, including teaching tips and other information alongside the lesson activities. You can also learn more about this lesson in my blog: *http://thelensthinkers.blogspot.com/search/label/Matter*.

| TABLE 3.1. Alignment of the Lesson to the *NGSS* | |
|---|---|
| **Connecting to the *NGSS*—Standard 5-PS1-3: Matter and Its Interactions** | |

*www.nextgenscience.org/pe/5-ps1-3-matter-and-its-interactions*

- The chart below makes one set of connections between the instruction outlined in this chapter and the *NGSS*.
- The materials, lessons, and activities outlined are just one step toward reaching the performance expectation listed below.

**Performance Expectation 5-PS1-3.** Make observations and measurements to identify materials based on their properties. (Examples of materials to be identified could include baking soda and other powders, metals, minerals, and liquids. Examples of properties could include color, hardness, reflectivity, electrical conductivity, thermal conductivity, response to magnetic forces, and solubility; density is not intended as an identifiable property. Assessment does not include density or distinguishing mass and weight.)

| Dimensions | Classroom Connections |
|---|---|
| **Science and Engineering Practices** | |
| *Analyzing and Interpreting Data*<br><br>Represent data in tables and/or various graphical displays (bar graphs, pictographs, and/or pie charts) to reveal patterns that indicate relationships.<br><br>Compare and contrast data collected by different groups in order to discuss similarities and differences in their findings. | Students analyze the patterns in their data to identify properties that distinguish solids, liquids, and gases. They use this as a basis to classify oobleck. |
| **Disciplinary Core Ideas** | |
| *PS1A: Structure and Properties of Matter*<br><br>Measurements of a variety of properties can be used to identify materials. | Students examine properties including mass, volume, shape, and more to compare different materials. They recognize which properties are most useful for identifying a material and which depend on the amount you have. |
| **Crosscutting Concepts** | |
| *Patterns*<br><br>Observed patterns in nature guide organization and classification and prompt questions about relationships and causes underlying them. | Students analyze patterns in their data to identify properties that distinguish matter from non-matter and to classify solids, liquids, and gases. This can be related to patterns important in other units such as phases of the moon and layers in rocks. |

Chapter 3

## Engage Phase (Day 1)

To begin this lesson, I asked my students to take a moment to think individually about the following:

**Teaching Tip:** Providing time for students to think and respond individually is important to learning, and it also can help the teacher document and assess the development of individual students' ideas as they work in groups.

- What the word "matter" means to them
- How they would explain what matter is to another person

These questions helped activate the students' background knowledge and provided them with the opportunity to consider the many meanings of matter.

After sampling a few students' responses, it became clear to the class that *matter* was a term that could be used in a variety of ways:

- "A *matter* of life and death"
- "It doesn't *matter*."
- "What's the *matter*?"

I explained to the students that scientists also use the word *matter* ("And anti-matter!" as one student called out) and that we were going to try to understand this meaning of matter better in our lesson. It's important to me that I help students recognize differences in the ways they may use words at home and in a scientific context.

**UDL Connection**
*Principle I: Representation*
*Guideline 3: Comprehension*
*Checkpoint 3.1: Activate or supply background knowledge*

The students were already placed in groups; however, before they worked as a group, I asked students to individually review a checklist and mark the items they believe are examples of matter. At the bottom of the checklist, they responded individually to the following question: "How did you decide whether something could be an example of matter? Explain your thinking."

After each group member had a chance to think and respond on his or her own, I provided the following guiding instructions and questions for the group discussion on the student recording sheet: "Compare your ideas with your group. What similarities and differences do you notice? Try to explain your thinking about why you did or did not check the items above."

As I circulated around the room, I encouraged students to jot down any additional notes or ideas from the group discussion that may have fit with their own thinking, contradicted their thinking, or created a question for further exploration.

**UDL Connection**

*Principle III: Engagement*
*Guideline 7: Recruiting Interest*
*Checkpoint 7.3: Minimize threats and distractions*

**UDL Connection**

*Principle II: Action and Expression*
*Guideline 6: Executive Functions*
*Checkpoint 6.4: Enhance capacity for monitoring progress*

Next, I asked students to work together to complete a card sort (see Figure 3.2, p. 56) that matches the list they just reviewed independently. The cards contained visuals of the words to support understanding across the languages for students who were English language learners (ELLs) and to make text accessible for my struggling readers. Students were to sort the cards collaboratively into two categories: *matter* and *not matter*.

**Teaching Tip:** Comparing a student's checklist with the group's card sort allows the teacher to ask the student to elaborate on his or her thinking. For example, "I noticed on the checklist that you marked that air is not matter. But when I look at your group's card sort, I noticed that *air* is in the matter category. Can you tell me about that change?"

**Figure 3.2.** Card Sort for the Engage Phase

| | | |
|---|---|---|
| water | cardboard | gravity |
| light | plastic | stone |
| oxygen | heat | sound |
| metal | oil | air |
| sand | | |

**UDL Connection**

*Principle I: Representation*
*Guideline 2: Language and Symbols*
*Checkpoint 2.4: Promote understanding across languages*

If your students are anything like mine, expect them to disagree on where some of the cards are placed. In this case, I suggested they might make a third pile. Many students in my class were conflicted with the cards labeled *sound*, *energy*, *air*, and *oxygen*. This seems developmentally appropriate as those examples are more abstract. If students cannot see or touch something, they can find it difficult to qualify it as matter.

---

**Teaching Tip:** One way to build interest and engage students in this phase of the lesson is to hold a gallery walk in which students can circulate around the room to view how their own ideas line up with those of other groups in terms of sorting the cards. When there are disagreements, this can prompt a desire to find the information—and it can be a great segue to exploration by emphasizing that when scientists disagree, they test their ideas against evidence.

---

To draw further on students' background knowledge, I asked each group to create a new card (using the blank spaces) with an example of their choice, and then as a class we discussed which group it belongs to. This assessment helped me identify any follow-up discussion that may be needed as I reviewed my plans for the next day.

## Explore Phase (Day 2)

During this phase, students had the opportunity to test out ideas. Working in their small groups as a way to keep those with behavior challenges focused and persisting in the lesson, the students rotated through a series of stations to explore the properties or characteristics of five items (light, air, water, stone, and sand) featured in the last activity. (Please see the "Materials and Safety Notes" box, p. 58.) Sand is included purposefully as one of the items because it forces students to clarify their thinking—their answers vary in whether they mean a *grain* of sand or the entire *bag* of sand—and it provides a way to differentiate the degree of complexity for the task to motivate some learners who need more of a challenge. This makes for great discussion in the next lesson phase!

**UDL Connection**
*Principle III: Engagement*
*Guideline 8: Sustaining Effort and Persistence*
*Checkpoint 8.2: Vary demands and resources to optimize challenge*

---

### Materials and Safety Notes

**Materials**

| | | |
|---|---|---|
| Water | Oil | Cups |
| Stone | Metal | Strainers |
| Sand | Plastic | Flashlights |
| Plastic tubs | Balloons | Oobleck in paper cups |
| Plastic bottles | Containers | Wax paper |

**Safety Notes**

1. Direct supervision is required during all aspects of this activity to ensure that safety behaviors are followed and enforced.
2. Make sure that any items dropped on the floor or ground are picked up immediately after working with them—a trip-and-fall hazard.
3. Immediately wipe up any water or other liquids spilled on the floor—a slip-and-fall hazard.
4. Never taste or drink any materials or substances used in the lab activity.
5. Wear nonlatex aprons when working with oobleck.
6. Use caution when working with glassware or plasticware. It can shatter and cut or scrape skin.
7. Keep all liquids away from electrical outlets to prevent shock.
8. Follow the teacher's instructions for disposing of waste materials.
9. Wash your hands with soap and water after completing this activity.

---

As students were exploring the materials in the station tubs (see the photos on the next page), they were asked to make sense of and answer each of the following questions on their student recording sheet:

1. Does it take up space?
2. Does it have mass?
3. Does it hold together? Can you pick up the whole thing?
4. Can you easily push another object or your hand through it?
5. Does it have a definite shape? (Does it keep the same shape when transferred to a different container?)
6. Does it have a definite volume? (Does it stay the same if you try to compress it?)
7. Can you pour it?

The students recorded their observations and ideas as they completed each of the seven activities as a tool for sharing and comparing ideas as a class. The stations do not have to be completed in any particular order, so the students are able to work at their own pace. To help those students who struggle with keeping organized and monitoring their progress, I provided data tables for them to record their findings.

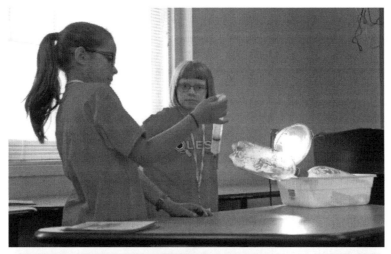

**Photos of students examining properties of matter in the Explore phase of the lesson. Having materials in tubs or containers that are clearly labeled allows groups to work independently as materials are organized.**

**UDL Connection**
*Principle II: Action and Expression*
*Guideline 6: Executive Functions*
*Checkpoint 6.3: Facilitate managing information and resources*

As the students circulated through the stations, I observed, documented, and anticipated what questions might benefit each group to help them with their understanding: "What are you noticing?" "How is _____ like _____?" I listened to the ways that students exchanged ideas, observed what vocabulary they used, and assessed what misconceptions they might have.

> **Teaching Tip:** Students might point out the expression "light *filled* the room," but that doesn't mean that light can fill a cup or take up space the way that water does.

## Explain Phase (Day 3)

The activity in this phase of the lesson was designed to help students clarify and refine their current understanding, first by talking with their group, and then by making sense of their data as a class with teacher support. Using the data collected in the Explore phase, the students completed a data chart to help them look for patterns and use the evidence to explain their thinking. I purposefully structured the data chart (see Figure 3.3) as a scaffold for data analysis, so that when students color-coded their responses, it allowed them to visually recognize patterns with greater ease.

Once each group filled in the chart and discussed their ideas, we compared the results—were there any disagreements? Sand provided a great topic for discussing how granular materials (such as sugar, sand, or salt) can have different properties when you consider questions such as "Can you pour it?" or "Can you pick up the whole thing?" in terms of a single grain or the entire amount you have. Students agreed that sand did not pour the same way that water did, so we marked that "No."

We revisited any other items that prompted ambiguous or inconsistent answers to reach a consensus about the properties of each. For example, students struggled with the idea of air having mass. In the Explore phase, the students tried filling a balloon with air and putting it on a digital scale and comparing the mass to an empty balloon. It often showed no difference. Comparing this to the weight of a single grain of sand as opposed to a bucket of sand can provide a useful analogy to make sense of it. The students realized that air is light enough that a small amount doesn't have much mass, just as a single grain of sand would not register on the digital scale.

**Figure 3.3.** Data Chart Used in the Explain Phase

| Question | Light | Air | Water | Stone | Sand |
|---|---|---|---|---|---|
| 1. Does it take up space? | | | | | |
| 2. Does it have mass? | | | | | |
| 3. Does it hold together in one piece when you pick it up? | | | | | |
| 4. Can you easily push something through it? | | | | | |
| 5. Does it have a definite shape? | | | | | |
| 6. Does it have a definite volume? | | | | | |
| 7. Can you pour it? | | | | | |

*Note:* The students color-code their "Yes"/"No" responses to identify patterns.

Next, I asked students to color-code the different responses in their data chart (e.g., coloring the "No" responses), to help the students who have difficulty identifying important or relevant ideas, as a way to highlight patterns and to help them recall these ideas for later tasks. Once the students reflected on their data chart, they were able to make the generalization that the first two characteristics (mass and volume) were shared by all items they considered to be matter. For light, which they considered *not* to be matter, they noted that they had not been able to demonstrate that light takes up space or has mass. This helped the class construct a formal definition of *matter* as something that both has mass and takes up space. Up until this point we had been using the phrase "taking up space." We clarified what that means (based on their evidence), and I indicated that this is referred to as *volume*. We also differentiated the use of the word to indicate loudness of a sound, and we linked the word to students' learning about area and volume in mathematics.

**UDL Connection**

*Principle I: Representation*
*Guideline 3: Comprehension*
*Checkpoint 3.2: Highlight patterns, critical features, big ideas, and relationships*

*Principle I: Representation*
*Guideline 3: Comprehension*
*Checkpoint 3.4: Maximize transfer and generalization*

## Extend Phase (Day 4)

In this phase, students extended their understanding by examining how properties of matter that *differed* between objects and materials might help us classify different types of matter further. Using the same core questions from the Explain phase, the students combined their data with new observations for oil, metal, plastic, and oxygen. Checking off the properties of each object listed on the chart, they were again looking for similarities or patterns in their data. Each small group was encouraged to reach an agreement and, if needed, go back to the materials in the previous lesson phase and retest them.

> **Teaching Tip:** Be sure to have enough testing materials for students to test the items on the list. For example, have containers, cups, strainers, a flashlight, and other items available that will help with the investigation.

Using the data from the chart, students devised a way to sort the items into new groups. The order in which I placed the items was intentional, so that similar types of matter were clustered together (see Figure 3.4). The students used color coding to indicate these groups. I asked them the following questions:

- Looking at the sections that are grouped similarly, what do you notice?
- What might you call each group?
- What other items do you think might belong in each group?

In my experience, with color coding, students could begin to see certain patterns emerge. Many quickly recognized that matter could be grouped into a solid, a liquid, or a gas. In some cases, students needed support and encouragement to recall what they observed in the Explore phase of the lesson, or I asked prompting questions such as "What one word could you use that would mean water, oil, or milk?" (*liquids*)

> **Teaching Tip:** As I work with students, I ask myself, "What misconceptions do my students hold, and what experience could I provide for them to further test their ideas?" I often challenge them with new examples or materials "on the fly," so it's helpful to have some extra materials on hand. I suggest salt, rubber, and aluminum foil.

Though students were familiar with the terms *solid*, *liquid*, and *gas*, this was a light-bulb moment for them in terms of how these classifications came about—and how they are defined by common properties. This

**Figure 3.4.** Data Chart From the Extend Phase of the Lesson

| Matter | It holds together in one piece | It is easy to push something through it | It has definite shape | It takes the shape of the container | It has definite volume |
|---|---|---|---|---|---|
| Oil | | | | | |
| Water | | | | | |
| Sand | | | | | |
| Stone | | | | | |
| Metal | | | | | |
| Plastic | | | | | |
| Air | | | | | |
| Oxygen | | | | | |

was also a light-bulb moment for me, as I would have typically introduced the classifications of solids, liquids, and gases first, then introduced students to their properties.

To wrap up the lesson, we constructed a class chart to help organize the information—and make explicit the critical ideas that some students need extra support in drawing out—as a way to solidify (pun intended!) the students' understanding (see Figure 3.5, p. 64).

**UDL Connection**

*Principle I: Representation*
*Guideline 3: Comprehension*
*Checkpoint 3.3: Guide information processing and visualization*

## Evaluate Phase (Day 5)

By this point in the learning cycle, I wanted my students to be able to use the properties of an object to identify it as matter and classify it as solid, liquid, or gas. Providing students with the opportunity to apply what they know to a new experience would give me information to determine if the students have a better conceptual understanding of what matter is as a result of my instruction and whether they can analyze and interpret their data to classify a substance as a solid, liquid, or gas.

---

**Figure 3.5.** Summary Chart Identifying Properties of Solids, Liquids, and Gases

**Properties shared by the group ...**

| | | |
|---|---|---|
| ➢ Holds together in one piece | ➢ Takes the shape of the container | ➢ Without a definite volume |
| ➢ Has a definite shape | ➢ Easy to push something through | ➢ The volume fills the whole container |

**Members of this group ...**

| | | |
|---|---|---|
| ➢ Stones, sand, metal, plastic | ➢ Oil, water | ➢ Oxygen, air |

**Name for this group ...**

| | | |
|---|---|---|
| ➢ Solid | ➢ Liquid | ➢ Gas |

In the final activity, I chose to present students with *oobleck*, a non-Newtonian substance that uniquely behaves as both a liquid and a solid—and it provides a great medium to challenge students' ideas about matter. While the version I use is a combination of cornstarch and water, a quick web search for "oobleck" will provide an assortment of recipes and variations for making your own. I prepared the oobleck ahead of time and distributed a portion of the "mystery substance" in a small paper cup for each student. I also provided each student with a sheet of wax paper as a placemat for easy cleanup. Before handing out these materials, however, I reviewed the exit slip on which students would be asked to individually record and self-assess their thoughts, to gauge and monitor their current understanding about matter. This review helped some of my learners better understand what they were monitoring, as a way to guide their own learning, because some of my students struggled to pose questions while completing tasks.

The exit slip that I used included the following questions:

- Is this substance matter? Provide evidence to support your claim.
- Can you classify this substance as a solid, liquid, or gas? Explain your reasoning.
- Can you think of something else that behaves like a liquid *and* a solid?
- What questions are you still wondering about? What is something that seems confusing or conflicting at this time?

For students who were struggling with a written response, I allowed them to either dictate their thinking to me as I scribed their responses or use voice-to-text on a tablet to support them as they were communicating their ideas.

**UDL Connection**

*Principle II: Action and Expression*
*Guideline 6: Executive Functions*
*Checkpoint 6.4: Enhance capacity for monitoring progress*

**UDL Connection**

*Principle II: Action and Expression*
*Guideline 5: Expression and Communication*
*Checkpoint 5.2: Use multiple tools for construction and composition*

As the students received their oobleck, there were many *oohs* and *ahhs*! Some immediately reached into their cups to touch it, while others attempted to pour it out onto the wax paper. As they explored, students noticed that it pours like a liquid but can be rolled in a ball and picked up in one piece like a solid.

For students who were hesitant to explore, I found action-testing questions (Elstgeest 1985) to be helpful. These "What happens when you ... ?" questions can only be answered by students if they try out the action and observe what happens. Examples include the following:

- "What happens when you jab or push on the oobleck slowly? Quickly?"
- "What happens when you squeeze the oobleck in your fist?"
- "What happens when you open your fist?"
- "What happens when you try to stir your oobleck?"

By asking these questions, I could ensure that my students were able to see the full range of properties of the oobleck to prepare them to answer the questions on the exit slip.

All students were able to justify that oobleck is matter, and most accurately identified that it has properties of both solids and liquids (though some were stuck with feeling that it needs to be one or the other). Other examples that students raised include pudding and gelatin—or things like chocolate that are solid at room temperature and behave like a liquid when the temperature increases. Asking students to share their ideas and state something that they were still unsure about, or something that did not make sense, opens the door for me to address their confusion in future lessons and experiences. These examples from students' exit slips can be integrated easily into future lessons about changes in matter!

## Unpacking UDL: Barriers and Solutions

I teach in a suburban school district that is located in a university town in the Midwest. I am one of five fifth-grade teachers at the school, which has 724 students—63% are White, 28% qualify for free or reduced lunch, 18% are ELLs, and approximately 7% have an individual education plan (IEP). I firmly believe that all of these students should have access to learning science. Although I had laid out this storyline, I knew that the activities could still pose potential barriers for my students. Using the Universal Design for Learning (UDL) framework and principles (described in Chapter 2), I was able to anticipate challenges that might arise and put specific strategies in place to provide a supportive learning environment for my students. For example, I knew that providing information in text alone would make things difficult for my ELLs. I also knew that for most of the students in my class with behavioral challenges, keeping focused on the task would be difficult.

Although I planned with specific students in mind, I've found that the UDL strategies I implement will benefit most students, regardless of individual learning strengths and challenges. Table 3.2 summarizes the UDL principles, guidelines, and checkpoints that I applied when designing the activities for each phase of the learning cycle lesson to meet the *general* needs of the learners in my classroom—specifically, organizing information, comprehending, staying focused and persisting in a task, and monitoring understanding of learning. Following the table, in the "learner profiles," I provide some examples of how I identified barriers and strategized solutions to meet the *specific* needs of two of the learners in my classroom.

| TABLE 3.2. Alignment of the Lesson With Principles of UDL | |
|---|---|
| **Connecting to the Principles of Universal Design for Learning** | |
| ***Principle I: Representation*** | |
| *Guideline 2: Language and Symbols* | |
| Checkpoint 2.1. Clarify vocabulary and symbols | During the English language arts lesson, students completed a Frayer Model on the word matter to support vocabulary development for ELLs. |
| Checkpoint 2.4. Promote understanding across languages | Pictures were added to words on the cards used during the card sort to support understanding for ELLs. |

*(continued)*

| **TABLE 3.2. Alignment of the Lesson With Principles of UDL** (*continued*) | |
|---|---|
| **Guideline 3: Comprehension** | |
| Checkpoint 3.1. Activate or supply background knowledge | Students discussed the different meanings for the word matter to activate and provide background knowledge for a shared understanding to complete the learning cycle. |
| Checkpoint 3.2. Highlight patterns, critical features, big ideas, and relationships | During the Explain phase, students were provided highlighters and a color-code key to highlight findings for identifying patterns. |
| Checkpoint 3.3. Guide information processing and visualization | During the Extend portion of the learning cycle, a class chart was created to help make ideas explicit and to solidify student understanding. |
| Checkpoint 3.4. Maximize transfer and generalization | During the Explain phase, the data were color-coded to facilitate the transfer of ideas about matter for the remaining phases of the learning cycle. |
| **Principle II: Action and Expression** | |
| **Guideline 5: Expression and Communication** | |
| Checkpoint 5.2. Use multiple tools for construction and composition | For students struggling with writing, they were provided different tools (dictate to a scribe or voice-to-text) to communicate ideas. |
| **Guideline 6: Executive Functions** | |
| Checkpoint 6.3. Facilitate managing information and resources | When collecting data during the Explore phase, students were provided data tables to manage and record findings. |
| Checkpoint 6.4. Enhance capacity for monitoring progress | In the Engage phase, a checklist with questions asked students to reflect on what they knew and as a way to promote group discussion. Additionally, before completing a task, students reviewed the questions that would be asked on the exit slip, to prompt self-assessment and understanding. |
| **Principle III: Engagement** | |
| **Guideline 7: Recruiting Interest** | |
| Checkpoint 7.3. Minimize threats and distractions | Students were provided a checklist to individually complete to ensure that they all had ideas to contribute to a group discussion. |
| **Guideline 8. Sustaining Effort and Persistence** | |
| Checkpoint 8.2. Vary demands and resources to optimize challenge | During the Explore phase, students rotated through stations that were designed to explore the properties of different items, including one item that would challenge thinking and vary the demands of the task. |

### Learner Profile: Derek

"Derek" was a student who received special education services in the learning disability category. He performed academically on a third-grade level in reading, writing, and math. His IEP included goals for developing executive-functioning skills and comprehension. I anticipated that the Explain phase would be particularly challenging for Derek, as there was a lot of information to make sense of from the Explore phase. Knowing that identifying patterns across the data could be a potential barrier, even though it was organized in a data table (to help students keep information organized and in mind), we planned the activity to include color-coding of the data. I took it one step further and added a "color-code key" to help students interpret the data.

Derek was given a piece of paper that had him use specific colors for "Yes" and "No." As a result, he did the following:

- Recorded all of the Ys for "Yes" first, with a green highlighter (green means *go/yes*), as he went across each row.
- Marked the empty spaces with an *N* for "No." He double-checked his data from the Explore phase to ensure that the empty spaces were really a "No" in his data. He then highlighted the No boxes with a pink marker (red means *stop/no*, but pink allows the student to still see the work and it is easily connected to the color red as they are similar in tone).

After Derek had color-coded his table in this way, he was told to focus on all of the green boxes as a way to identify and see the pattern across the data table. While this was planned with Derek in mind, I noticed that his group members were also struggling with identifying patterns when the table initially contained just "Yes" or "No." Once all students had visually coded their data, they were able to identify a pattern and explain that the properties that were green (mass and volume) were unique to matter.

### Learner Profile: Thanh

At my school, we commonly encounter students from other countries who are visiting in our city because they have family members doing research at the local university. "Thanh" was proficient in her native language, and she had a working understanding of English. However, she struggled with content-specific academic language.

In the first phase of the learning cycle (Engage), I incorporated pictures along with the words on the cards, so that students could make sense of any unfamiliar objects. For Thanh, I also anticipated that using the term *matter* in a scientific sense could be a barrier—particularly because in English we use the word in many nonscientific ways. I decided to acknowledge this at the beginning of the Engage phase, to signal to students that a familiar word was going to be used in a specific way for a specific purpose.

To further help students define *matter* in a scientific context and support their science learning in other content areas, I introduced the Frayer Model (Frayer, Frederick, and Klausmeier 1969; see Figure 3.6) during English language arts time as we reviewed vocabulary that was being used in the different assigned readings. We constructed this on chart paper and kept it posted in the room as a reference. Adding picture images to the class chart provided another layer of support for ELLs to help them not only make meaning, but also remember these new words in the context that they were learned. This tool was helpful not only in this lesson, but also throughout the unit to define academic vocabulary, including *gas, evaporation, condensation, chemical, property,* and other abstract words. Thanh was not the only student who referenced these charts during the lesson—other students also found them helpful when writing in their science notebooks and explaining their ideas to their peers.

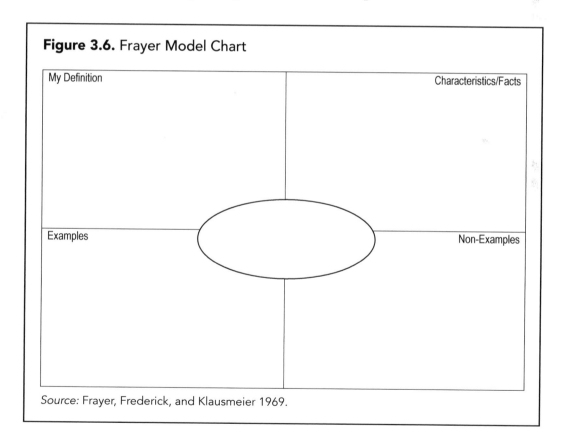

**Figure 3.6.** Frayer Model Chart

My Definition

Characteristics/Facts

Examples

Non-Examples

*Source:* Frayer, Frederick, and Klausmeier 1969.

## Supporting All Learners: Working in Groups

My students often work in groups. While this can be a source of support for some learners, I also have noticed at times that this can be a barrier as well. Sometimes, certain students will share their ideas immediately—before others have had a chance to think about the question or problem for themselves. As a result, the other students in the group might shut down their own thinking to defer to the authority of the student who is a "quick thinker." These same students, while they mean well, can dominate group discussions, often limiting the contributions of other group members.

In the Engage phase of the lesson, the students were placed in groups to complete a picture card sort activity. Immediately before that activity, I embedded a solution (see Table 3.2, p. 66) that was designed with specific learners in mind. The solution was designed to provide thinking time and facilitate active contribution to the discussion from all students in the group.

Specifically, in the Engage phase of the lesson, I first had the students individually complete a checklist with their own ideas about what is or isn't "matter"—this allowed all students to consider their own prior knowledge and experience before sharing with their group. After the students responded to the checklist, I provided instructions and questions for them to use to guide them in their group discussion. It also helped highlight any areas of disagreement in ideas among the members of the group—which I've found can be a motivator for students to find out what is correct.

While my students sometimes have asked me whether an answer was correct, in this lesson I really tried to empower the students to figure it out for themselves by testing their ideas with evidence. It's important to me that they don't accept answers based on authority, but that they learn to question ideas that aren't supported by evidence.

## Questions to Consider

➢ To what extent did the activities that this teacher chose align with the purpose and intent of each phase of the 5E Learning Cycle?

➢ Could you envision other activities that would be appropriate for each phase?

➢ Were you able to follow the sequence of activities and the ideas students developed in the learning cycle that the teacher created?

➢ How did the storyline of the lesson progress?

➢ In what ways was the teacher able to assess students during each phase of the lesson? How did this inform her instruction?

➢ As you read through this lesson, did you come across any activities that might pose a barrier for your own students? What principles of UDL might you apply in those instances?

## References

Elstgeest, J. 1985. The right question at the right time. In *Primary science: Taking the plunge*, ed. W. Harlen, 36–47. Oxford, England: Heinemann Educational Books.

Frayer, D., W. C. Frederick, and H. J. Klausmeier. 1969. *A schema for testing the level of cognitive mastery*. Madison, WI: Wisconsin Center for Education Research.

NGSS Lead States. 2013. *Next Generation Science Standards: For states, by states*. Washington, DC: National Academies Press. *www.nextgenscience.org/next-generation-science-standards*.

Wiggins, G., and J. McTighe. 2005. *Understanding by design*. 2nd ed. Alexandria, VA: Association for Supervision and Curriculum Development.

# CHAPTER 4

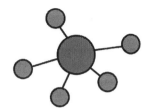

# To Change or
# Not to Change?

**Christine Meredith**

This chapter focuses on a lesson about phenomena related to chemical changes, in which new substances form when two substances are combined. This lesson appears in the middle of a unit on matter, when students have already learned about physical changes. They also have learned to identify effects that indicate when a chemical change has occurred (such as a gas being produced or heat being taken in or given off). The conceptual storyline for this lesson (see Figure 4.1, p. 74) builds from students' prior instruction on physical changes and is grounded in personally relevant examples from the students' daily life experiences.

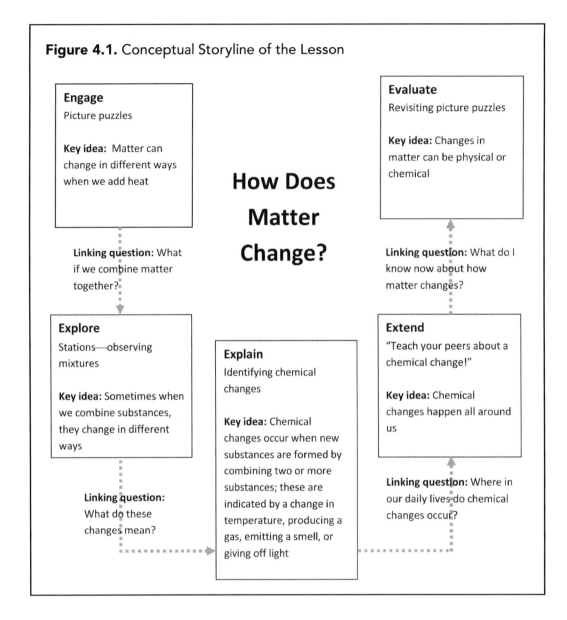

Figure 4.1. Conceptual Storyline of the Lesson

**How Does Matter Change?**

**Engage**
Picture puzzles

**Key idea:** Matter can change in different ways when we add heat

**Linking question:** What if we combine matter together?

**Explore**
Stations—observing mixtures

**Key idea:** Sometimes when we combine substances, they change in different ways

**Linking question:** What do these changes mean?

**Explain**
Identifying chemical changes

**Key idea:** Chemical changes occur when new substances are formed by combining two or more substances; these are indicated by a change in temperature, producing a gas, emitting a smell, or giving off light

**Evaluate**
Revisiting picture puzzles

**Key idea:** Changes in matter can be physical or chemical

**Linking question:** What do I know now about how matter changes?

**Extend**
"Teach your peers about a chemical change!"

**Key idea:** Chemical changes happen all around us

**Linking question:** Where in our daily lives do chemical changes occur?

## LESSON VIGNETTE

Matter may seem like a simple and straightforward topic—however, that's simply not true! Matter is complex. It can take different shapes and forms, and it can change: Add some heat energy and a solid becomes a liquid; combine two materials and you might get a totally new substance. As teachers, how can we help our students understand these kinds of changes?

Teaching about changes in matter has been a challenge for me since the beginning of my teaching career. My district really didn't have a current curriculum for science that aligned with the *Next Generation Science Standards* (*NGSS*), so I found I was relying on resources I located on the internet or on things I would come up with on my own. But those things didn't fully address the performance expectations or keep my students engaged. Then I attended the Quality Elementary Science Teaching (QuEST) workshop.

Participating along with other fifth-grade teachers in QuEST gave me a new desire and understanding of science and showed me how it should be presented to students. I was taught about the 5E Instructional Model as well as Universal Design for Learning (UDL). These were two components that I desperately needed to be successful in teaching science! The 5E model offered a way to organize my lessons, and UDL showed me how to adapt the resources and activities that I used to better support my students in learning at their own pace and in a way that made sense to them.

You could say I am no longer adrift. I have found my oars and have made headway to a better understanding of effective science teaching and engagement for all learners in my classroom. I am excited to share this lesson with you, and I hope it helps you on your own journey!

See Table 4.1 (p. 76) for alignment of this lesson to the *NGSS* (NGSS Lead States 2013).

| TABLE 4.1. *NGSS* Alignment | |
|---|---|
| **Connecting to the *NGSS*—Standard 5-PS1-4: Matter and Its Interactions** | |
| *www.nextgenscience.org/pe/5-ps1-4-matter-and-its-interactions* | |
| • The chart below makes one set of connections between the instruction outlined in this chapter and the *NGSS*.<br>• The materials, lessons, and activities outlined are just one step toward reaching the performance expectation listed below. | |
| **Performance Expectation 5-PS1-4.** Conduct an investigation to determine whether the mixing of two or more substances results in new substances. | |
| **Dimensions** | **Classroom Connections** |
| *Science and Engineering Practices* | |
| *Planning and Carrying Out Investigations*<br><br>Make observations and measurements to produce data to serve as the basis for evidence for an explanation of a phenomenon. | Students observe and record data about various mixtures before and after combining (temperature, color, thickness, etc.) to understand whether a chemical change has occurred and a new substance has formed. |
| *Disciplinary Core Ideas* | |
| *PS1.B: Chemical Reactions*<br><br>When two or more different substances are mixed, a new substance with different properties may be formed. | Students examine changes that occur when two or more substances combine, and they identify any new substances that are formed in the process. |
| *Crosscutting Concepts* | |
| *Cause and Effect*<br><br>Cause-and-effect relationships are routinely identified and used to explain change. | Students examine the effect of adding heat and of combining substances. This can be related to other cause-and-effect relationships they have encountered in previous units. |

## Engage Phase

I begin all my lessons for science with a flair. I put on music and shout, "Hey everyone, let's get ready to do sciiiiiiiiiiiiiiiience!"—similar to the well-known "Let's get ready to rumble!" that you might hear at sporting events. The students get up and jump around excitedly for science to begin. When the music stops, they know it's time to take their seats and get ready to listen to directions. During the first phase of the lesson, it's important to activate students' prior knowledge (what experiences have they already had that will influence how they think and understand chemical changes?) and identify any potential misconceptions they hold (what ideas have they formed based on their experiences?).

For this lesson, rather than simply asking my students a question, I projected several pictures on our smartboard, including a melting snowman, a fire in a fireplace, a kettle steaming, and an egg frying. (Several students laughed and pointed out how the snowman was melting, relating this to our own recent snow-melt.) I explained to the students, "I would like you to determine what these pictures are about and what changes are taking place. Think about what we learned previously about physical changes. Do you think these are all examples of that?"

I gave the students thinking time and then asked them to turn to their shoulder partner. The students were so excited that they were almost talking over each other. As they were sharing, I moved around listening to their conversations to hear what types of connections they were able to make:

Warren: *I built a snowman once, and I was really sad that it had melted.*

Tyler: *So how do you think the pictures are the same?*

Warren: *They are all changing.*

Tyler: *I think they are all changing because of heat.*

I was very pleased to see that Warren and Tyler were guiding their own conversation because this was a skill that we had been working on consistently since the beginning of the school year. This skill has really helped my struggling learners attend to a task as well as formulate an answer with guidance from a peer.

Students described the changes in the snowman as melting, the fire as burning, the egg as cooking, and the kettle as boiling—and like Tyler had suggested, they attributed these changes to heat. Most recognized melting as an example of a physical change by drawing on their knowledge from previous lessons, but the boiling kettle, frying egg, and burning wood were a little more puzzling to them. Some pointed out that you could refreeze the melted water, but that you couldn't unburn wood or unfry an egg. Still others suggested that the boiling water was still in the air (but were unsure whether you could get it back). We reached a consensus that some changes could be "undone" while others could not.

"Would it be possible to cause a change without adding heat?" I asked. Students turned back to their partners and drew on their prior experiences to brainstorm. When I called them back together, some suggested taking heat away or cooling something could cause it to change (such as freezing ice cubes), but students struggled to generate other examples. "What about if I were to combine two different things?" I suggested, asking students to talk again with their partners.

This time, students generated some fictional examples (mixing potions, like a wizard) and drew on food-related experiences (adding pancake mix and water). This discussion prompted them to want to learn more—several asked if they would get to mix things together today—so I knew I had them "hooked" and was ready to move on.

## Explore Phase

I told students that I had prepared some stations for them to explore to answer the question, "How does matter change if we combine two things?" The four stations I prepared were as follows:

- Station 1—borax and white school glue
- Station 2—salt and water
- Station 3—lemon juice and milk (warm or room temperature)
- Station 4—sodium bicarbonate (such as Alka-Seltzer) and water

Please see the "Materials and Safety Notes" box below.

---

### Materials and Safety Notes

**Materials**

| | | |
|---|---|---|
| Borax | Lemon juice | Stir sticks |
| White school glue | Milk | Thermometers |
| Salt | Sodium bicarbonate | Empty water bottle |
| Water | Paper cups | Balloon |

**Safety Notes**

1. All involved must wear indirectly vented chemical splash goggles during all phases of these inquiry activities (setup, hands-on investigation, and take-down).
2. Direct supervision is required during all aspects of this activity to ensure that safety behaviors are followed and enforced.
3. Make sure that any items dropped on the floor or ground are picked up immediately after working with them—a trip-and-fall hazard.
4. Immediately wipe up any water or other liquids spilled on the floor—a slip-and-fall hazard.
5. Never taste or drink any materials or substances used in the lab activity.
6. Wear nonlatex gloves when working with borax and avoid situations where skin might be in contact with the plaster.
7. Use caution when working with glass or plasticware. It can shatter and cut or scrape skin.
8. Keep all liquids away from electrical outlets to prevent shock.
9. Follow the teacher's instructions for disposing of waste materials.
10. Wash your hands with soap and water after completing this activity.

---

Each work area had the substances premeasured into individually labeled paper cups, along with trays and other materials such as stir sticks and non-mercury thermometers. Although most students did not have difficulty with measuring, my student Betsy had limited motor skills. I wanted to ensure that she could fully participate along with anyone else who might have difficulty with measuring.

I previewed these stations for students, asking them what kinds of "changes" they might expect to observe. Some students suggested that changes in color might occur, while others proposed that bubbles might form. "During your exploration, I would like you to observe and record your findings on the recording sheet that I've provided," I told the class. See Figure 4.2 for a sample recording sheet.

**Figure 4.2.** Sample Recording Sheet for Exploration Stations

| Station 1: Borax and Glue | Station 2: Salt and Water |
|---|---|
| Observations of borax: | Observations of salt: |
| Observations of glue: | Observations of water: |
| Observations of mixture: | Observations of mixture: |
| • Does it cause a smell?<br>• Does it give off a gas?<br>• Does it give off light?<br>• Does it change temperature?<br>• Does it change color?<br>• Does it change thickness? | • Does it cause a smell?<br>• Does it give off a gas?<br>• Does it give off light?<br>• Does it change temperature?<br>• Does it change color?<br>• Does it change thickness? |
| **Station 3: Lemon Juice and Milk** | **Station 4: Sodium Bicarbonate and Water** |
| Observations of lemon juice: | Observations of sodium bicarbonate: |
| Observations of milk: | Observations of water: |
| Observations of mixture: | Observations of mixture: |
| • Does it cause a smell?<br>• Does it give off a gas?<br>• Does it give off light?<br>• Does it change temperature?<br>• Does it change color?<br>• Does it change thickness? | • Does it cause a smell?<br>• Does it give off a gas?<br>• Does it give off light?<br>• Does it change temperature?<br>• Does it change color?<br>• Does it change thickness? |

**UDL Connection**

*Principle II: Action and Expression*
*Guideline 5: Expression and Communication*
*Checkpoint 5.3: Build fluencies with graduated levels of support for practice and performance*

I reviewed each of the questions on the sheet with the students:

- Does it cause a smell?
- Does it give off a gas?
- Does it give off light?

- Does it change temperature?
- Does it change color?
- Does it change thickness?

I also suggested other types of changes that might be important to note, in addition to what the students had suggested.

I explained to the students that there were enough materials so that each pair could use one paper cup of each substance to combine. Students worked in pairs for this phase and other phases of the lesson as a way to support and encourage one another as needed (e.g., when a student needed help moving and manipulating materials or completing tasks involving reading and writing). Students and their partners could begin at any station they wanted and could proceed in any order. We reviewed our safety rules (see the box, p. 78), and I paused to answer any questions before allowing the students to begin.

**UDL Connection**

*Principle III: Engagement*
*Guideline 8: Sustaining Effort and Persistence*
*Checkpoint 8.3: Foster collaboration and community*

The students immediately started talking and hurrying to their chosen station. As I walked around, I noticed that many students were discussing how they would combine the two substances, that is, which they would pour into the other. I was happy to see that they were formulating a plan in order to begin their work.

As I moved around the room, I found that Betsy and her partner Baylee were having trouble getting Betsy (who had limited mobility) close enough to the table for her to be able to participate in the activity effectively. I went up to the girls to discuss their dilemma:

Teacher: *Have you ladies started your exploration?*

Betsy: *I can't reach the items on the table.*

Teacher: *So, how can we solve that problem? (The girls stared at me for a moment.)*

Baylee: *We could take a tray of materials back to our desks and then we could do the experiment.*

Betsy: *Then I can reach the items since my chair fits under my desk.*

Teacher: *I think that's a great idea.*

So, the girls took the items that were needed and went back to their seats to work on the experiment together.

As I continued to walk around the room, I could see that the students were excited because they were smiling and talking about what was happening to the different materials that were being combined. As they were filling out the recording sheet, however, I saw that they were only putting one-word comments. So then I asked the class to "Gimme 5"—our usual attention grabber. The students quickly raised their hands and turned off their voices so they could hear what the new instruction would be. When all of the students had stopped talking and were ready to listen, I asked, "What kind of responses do I expect on your recording sheets?"

"A complete thought or picture with labels!" Warren recommended.

"That's correct. You may choose how to answer, but make sure you are giving enough detail so when you look back at your notes you will be able to tell everyone else what happened after you combined the two substances," I reminded them.

With that simple reminder, I immediately saw students start to record their observations with more details. The students continued their investigations and finished recording their findings. As we cleaned up for the day, I knew that students would be ready to move on to the next phase of the lesson tomorrow.

**UDL Connection**
*Principle II: Action and Expression*
*Guideline 5: Expression and Communication*
*Guideline 5.1: Use multiple media for communication*

**Teaching Tip:** For Station 1, I encourage students to remove the "slime" from the cup and to observe it through touch. This way, the students are able to observe that it feels cool to the touch. I then ask them to return the slime to their cup and wash their hands before continuing to any remaining stations. As an option, you might place the slime in a resealable plastic bag for students to take home.

## Explain Phase

I gathered the whole class together the next day and asked, "Who can remind me what happened yesterday?" Immediately, hands shot up.

"We combined things together to see how they changed," Tyler offered.

"And how did things change?" I probed further. Immediately, the sound of recording sheets being pulled out of notebooks filled the room. "Aha! Our recording sheets would be helpful now!" I said, smiling. It's important to me that students' writing is for a purpose, and using it in our discussion helps demonstrate that.

I gave the students time to retrieve their recording sheets and then asked them to do a "round robin," in which they share by going around in a circle so that all pairs have the opportunity to communicate what they observed and experienced during their exploration. Using the document camera, I completed a class recording sheet as we reached consensus about the kinds of changes that occurred when substances were combined at the four stations.

**Teaching Tip:** In this particular case, all of our observations were consistent. However, in the event that groups had differences in observations, it's a good idea to have extra materials on hand to retest and resolve inconsistencies.

The students connected the changes in Station 2 to our earlier focus on physical changes, but they didn't think that the other changes they observed were physical changes, but something else. To support students in making sense of their observations, I selected a video from Crash Course Kids (see "Resources") that describes chemical changes. I chose the video as a way to help my students connect what they had experienced in order to clarify vocabulary in a way that was accessible for all learners. I told the students the following:

*Now that you've completed your explorations, I would like you to think about the things that you observed at Stations 1, 3, and 4. What do these changes mean? We're going to watch a video that can help us make sense of our observations. I would like you to pay close attention to the definition of a chemical change, and how you can tell whether you've observed a chemical change taking place.*

**UDL Connection**
*Principle I: Representation*
*Guideline 2: Language and Symbols*
*Checkpoint 2.1: Clarify vocabulary and symbols*

I wrote two sentence openers on the board, indicating that I wanted the class to help me complete them after the video:

1. "A chemical change is when …"
2. "You know a chemical change has occurred if …"

After the video, students immediately connected the ideas to the questions I had them include on their recording sheets! We completed the sentences as follows:

1. "A chemical change is when *two or more substances combine and the particles rearrange to form a new substance.*"
2. "You know a chemical change has occurred if *it makes a smell, gives off gas, releases light, or gives or takes heat.*"

I asked the students to go back to the evidence to make a claim about which combinations, if any, were examples of chemical changes. I provided a new recording sheet on which students could develop their arguments (see Figure 4.3).

---

**Figure 4.3.** Recording Sheet for the Explain Phase

**Properties of substances before combining:**

Substance 1:

Substance 2:

**Changes observed when combining the substances:**

**Properties of the mixture:**

Claim: I think a _____ change occurred.

**Evidence:**

Reasoning: This supports my claim, because in this type of change _____

_____

_____

---

For Station 1, the class reached a consensus that the mixture of borax and glue resulted in a chemical change. Not only did they recognize that the thick and gooey result seemed to be a new substance, but they also reported that the temperature dropped and the "slime" substance felt cool to the touch. I have to clarify that it was both cool in the sense of awesome and cool in the sense of not being warm.

For Station 3, lemon juice and milk, the students also were confident that a new substance had been formed—the "curdles" of milk are not something students can easily identify, so there were all sorts of descriptions from "puke-like" to "chunky stuff." I asked the students whether they had ever tried cottage cheese or heard the nursery rhyme about Little Miss Muffet eating her "curds and whey," and explained that these are the same thing and are what they saw forming when they combined milk and lemon juice.

For Station 4, sodium bicarbonate and water, a majority of students claimed that it was a chemical change because a gas was produced ("There were bubbles!"). However, some pointed out that it was fizzy like the soda they drink, which also has bubbles, and that there are also bubbles in their bathtub when they use bubble bath. They were confused about whether these are all the same.

Luckily, I had extra supplies on hand, as well as an empty water bottle and balloon. I offered to do a demonstration so that we could visualize the gas being produced when we combine the two. I placed water in the bottle, dropped in a sodium bicarbonate tablet, and quickly sealed the bottle with the balloon. Slowly, the balloon began to inflate. While this was enough to convince some students that the gas was a new substance that wasn't there before, others remained skeptical. Rather than correct students or tell them the "correct" answer, I suggested that perhaps we needed to explore this particular idea further and gather more evidence in our upcoming lessons.

## Extend Phase

An important part of this phase of the learning cycle is helping students apply their new ideas to a new context. I wanted this next part of the lesson to be motivating for students, and to position them as "experts" to teach one another. I also wanted it to be something that helped the students connect the ideas they were learning to the world outside our classroom. While teachers often present phenomena to students as part of *NGSS*-aligned science instruction, I decided to ask the *students* to identify and present phenomena to one another! I proposed the following to them:

> *Let's see what you have learned about chemical changes. I want you to identify a chemical change that takes place in your everyday life. I want you to work in pairs to set up an exploration demonstration for the class to see or experience this change, and then be able to explain what makes it a chemical change. We will present these in one week! You should use everyday types of matter within your example that you know you can combine safely. Once you have come up with the examples that you would like to use, fill out the planning checklist and please bring it to me for approval.*

**UDL Connection**
*Principle III: Engagement*
*Guideline 7: Recruiting Interest*
*Checkpoint 7.2: Optimize relevance, value, and authenticity*

**UDL Connection**
*Principle II: Action and Expression*
*Guideline 6: Executive Functions*
*Checkpoint 6.2: Support planning and strategy development*

I could tell that students were extremely excited about this task, and eager to get started—so I knew the planning sheet would be an important scaffold to ensure they were thinking through their plans carefully. Each day leading up to the presentations, I provided some time in class for students to work on their plans and to check in with me on their progress. By doing so, I could provide conceptual support when students were struggling to identify scientifically accurate examples of phenomena in which a chemical change occurs or to form explanations based on evidence. This also allowed me to ensure that I had all the materials on hand that students would need, and it gave them the chance to try out their ideas before presenting.

> **Teaching Tip:** Although I provide all materials for students, occasionally there are things they want to bring from home. I include a note about this assignment in our class newsletter to give parents and guardians a "heads up."

> **Teaching Tip:** This particular task would go well with helping students meet learning goals in English language arts as well—and it will allow you to combine standards from both areas!

On the day of the presentations, the students entered the classroom visibly excited for the day! I knew it would be all they could focus on, so I opted to start the day off with science (which I usually taught in the afternoons). Before we began, I reminded the class about appropriate audience behavior and listening skills—as well as how to ask questions of the presenters if anything was unclear.

## Evaluate Phase

Students' presentations will serve as one form of summative assessment of their learning, specifically their grasp of the disciplinary core idea that *when two or more different substances are mixed, a new substance with different properties may be formed.* Additionally, to help students reflect on what they have learned and how their ideas have changed, I ended our lesson by reposting the four pictures we initially viewed in the Engage phase and asking them to individually classify each as a chemical or physical change. As I reviewed students' answers, I saw that many were even more confident that the melting snowman was a physical change, and most were able to correctly identify burning wood and a frying egg as chemical changes—both because the burning wood was giving off light and because the frying egg was a new substance with different properties from a raw egg.

However, I recognized one idea that some students were having trouble integrating into their understanding—the idea that chemical changes involve giving

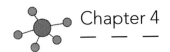

off or taking in heat. Some were confused about whether *adding heat* in the four scenarios is the same as *taking in heat* in a chemical reaction. I was not discouraged, because this reflects students' efforts to connect two things they've learned, but I recognized that I had an important job to do in helping them resolve this confusion in our next lessons, along with clarifying whether a gas was being produced!

## Unpacking UDL: Barriers and Solutions

It's important to me that all students feel successful in learning science. To ensure that, I have to make sure that I set them up for success—by planning lessons that don't unintentionally pose barriers for them. Table 4.2 summarizes the UDL principles, guidelines, and checkpoints that I applied when designing the activities for each phase of the learning cycle lesson to meet the general needs of the learners in my classroom. Following that are some examples of how I identified barriers and strategized solutions to meet the specific needs of two of the learners in my classroom.

| TABLE 4.2. UDL Connections | |
|---|---|
| **Connecting to the Principles of Universal Design for Learning** | |
| **Principle I: Representation** | |
| *Guideline 2: Language and Symbols* | |
| Checkpoint 2.1. Clarify vocabulary and symbols | A video was used to provide information and clarification of key concepts in a format that was accessible for all students. |
| Checkpoint 2.5. Illustrate through multiple media | Pictures on worksheets were provided instead or alongside words on a worksheet for Warren. |
| **Principle II. Action and Expression** | |
| *Guideline 4: Physical Action* | |
| Checkpoint 4.1. Vary the methods for response and navigation | Materials were organized on paper plates or in an accessible location for Betsy to be able to interact with hands-on components of the lesson. |
| Checkpoint 4.2. Optimize access to tools and assistive technologies | An alternative keyboard with large keys was readily available with a computer for Betsy to use where needed. |

*(continued)*

| TABLE 4.2. UDL Connections (*continued*) | |
|---|---|
| **Guideline 5: Expression and Communication** | |
| Checkpoint 5.1. Use multiple media for communication | Students in the Explore phase could record ideas via either words or pictures to communicate their ideas. |
| Checkpoint 5.2. Use multiple tools for construction and composition | For Betsy, she used talk-to-text software as a way to record her ideas. |
| Checkpoint 5.3. Build fluencies with graduated levels of support for practice and performance | In the Explore phase, materials to be used in the stations were premeasured to support a student with limited mobility and other students who had difficulty with measuring. |
| **Guideline 6: Executive Functions** | |
| Checkpoint 6.2. Support planning and strategy development | A planning form (checklist) was provided to help students with recalling what to do to help complete a task. |
| **Principle III. Engagement** | |
| **Guideline 7: Recruiting Interest** | |
| Checkpoint 7.2. Optimize relevance, value, and authenticity | For the Extend phase, to motivate and help connect ideas to a real-world situation, and the challenges that some students experience, the students had to identify chemical changes that occur in their everyday life. |
| **Guideline 8. Sustaining Effort and Persistence** | |
| Checkpoint 8.3. Foster collaboration and community | Students in various phases were placed in pairs to provide each other with additional support such as manipulating materials or help with writing and reading. |

### Learner Profile: Betsy

"Betsy" was a student who had been diagnosed with cerebral palsy. She spent the majority of her day in a wheelchair and had very limited fine motor skills. However, Betsy was very intelligent and enjoyed learning right beside her peers. Throughout this lesson, there needed to be several changes made for Betsy to be successful in her discoveries and learning. Some of the solutions I used were unique to her needs and connected to her individual education plan accommodations and modifications. She used them at any point throughout the lesson where needed, and they also reflected UDL solutions. Our solutions included the following:

- Talk-to-text to allow her to record her thinking and information
- An alternative keyboard with bigger keys to type with

- Alternative ways to physically interact with equipment (e.g., taking materials back to her desk, putting materials on paper plates as a way to get closer and interact with them)

**UDL Connection**

*Principle II: Action and Expression*
*Guideline 4: Physical Action*
*Checkpoint 4.1: Vary the methods for response and navigation*

**UDL Connection**

*Principle II: Action and Expression*
*Guideline 4: Physical Action*
*Checkpoint 4.2: Optimize access to tools and assistive technology*

**UDL Connection**

*Principle II: Action and Expression*
*Guideline 5: Expression and Communication*
*Checkpoint 5.2: Use multiple tools for construction and composition*

I embedded two solutions in the lesson with Betsy in mind:

- Betsy had a hard time using her hands to manipulate different items. Because this was difficult for her, I had all of the materials already measured out and put into larger containers for her to put together. This really helped her be able to get the matter from one container to another without spilling or dropping the materials.
- I had Betsy working with a partner to help her get items into containers that were more manageable for her to perform her explorations and identify the correct change and evidence to support her claim.

These two solutions were also made available for all students and were of benefit for others.

### *Learner Profile: Warren*

"Warren" was a student with an intellectual disability. He had an IQ of 55 and came into my classroom from a self-contained room. Warren was very energetic and loved to talk, often needing redirection or prompts to stay on topic. In teaching any lesson with Warren, there were a number of accommodations and modifications that he received. He could use them at any point throughout the lesson where needed. These also reflected UDL solutions.

For this lesson, Warren's modifications included a worksheet with pictures instead of words (or with words) to make connections to the explorations that took place earlier in the lesson. Warren was able to match the pictures that he drew earlier in the lesson. With the help of the pictures, he was able to see how the pictures he drew in the Explore phase helped him understand that there were changes taking place and which he thought were chemical and physical changes.

**UDL Connection**

*Principle I: Representation*
*Guideline 2: Language and Symbols*
*Checkpoint 2.5: Illustrate through multiple media*

I also built in several UDL solutions that were designed with Warren (and others) in mind as a way to increase his access to the learning:

- Instead of writing, which he found very difficult, Warren was allowed to draw pictures and, if possible, label the things that he needed so he would remember what he saw during the different explorations. He made sure that he got help from his partner to ensure that we could read what he was trying to tell us through his picture.

- Warren was paired with another student from my class who enjoyed working with him and was able to help him with reading, writing, and any math that he was required to do to complete the lesson successfully. The partner often helped guide and lead him in the right direction for his project.

- Warren was also given a checklist and organizers to help him be better prepared and to give him a step-by-step list to help him know what he should be doing.

---

## Questions to Consider

➢ To what extent did the activities in this lesson align with the purpose and intent of each phase of the 5E Learning Cycle?

➢ Could you envision other activities that would be appropriate for each phase?

➢ Were you able to follow the sequence of activities and the ideas that students developed in the lesson?

➢ How did the storyline of the lesson progress? How did the teacher begin thinking about how to connect to the storyline of future lessons?

➢ In what ways was the teacher able to assess students during each phase of the lesson? How did this inform future instruction?

➢ In what ways did the solutions that the teacher identified meet the needs of the specific students spotlighted in this vignette? In what ways did they benefit all students? Could you think of other solutions you might use for your learners?

---

## Reference

NGSS Lead States. 2013. *Next Generation Science Standards: For states, by states.* Washington, DC: National Academies Press. *www.nextgenscience.org/next-generation-science-standards.*

## Resources

American Chemical Society. "Time for Slime!" Chemistry for Life. *www.acs.org/content/acs/en/education/whatischemistry/adventures-in-chemistry/experiments/slime.html.*

Crash Course Kids. "Chemical Changes." Episode 19.2. *www.youtube.com/watch?v=37pir0ej_SE.*

# CHAPTER 5

# Magnets— What's the Problem?

**Deborah Hanuscin**

In this chapter's lesson, third graders explore how magnets interact with each other and with other materials, and then they analyze the magnetic forces and interactions being used in the design of various products and tools to solve problems. This lesson serves as an introduction to a unit that focuses on designing a magnetic product or invention to solve a problem. Universal Design for Learning (UDL) principles are applied to meet the needs of struggling readers and writers. The conceptual storyline for the lesson (see Figure 5.1, p. 92) begins by presenting students with a series of claims about magnets, the testing of which serves to guide their learning.

**Figure 5.1.** Conceptual Storyline of the Lesson

## How Can We Use Magnets to Solve Problems?

**Engage**
Agree/disagree statements

**Key idea:** When there are conflicting claims, they should be checked against evidence

**Linking question:**
How can we test our ideas?

**Explore**
Hands-on investigation of magnetic interactions

**Key idea:** Magnets interact in different ways

**Linking question:**
What ideas about magnets are correct?

**Explain**
Revising claims to be consistent with evidence

**Key idea:** Magnets attract some metals, attract and repel each other, and can interact both through other materials and at a distance

**Evaluate**
Analyzing and explaining a magnetic solution to a problem

**Key idea:** Scientific ideas can be applied to solve problems

**Linking question:** How can magnets be used to solve problems?

**Extend**
Reverse-engineering magnetic products

**Key idea:** Magnets can be used to solve problems

**Linking question:** How do we use magnets in our everyday lives?

## LESSON VIGNETTE

I first taught magnets as an elementary teacher, and later as a museum educator, a workshop facilitator, and finally as a professor of science education and physics. Needless to say, my practice has evolved during that time! While I began teaching about magnets at first, I have transitioned to using magnets as a vehicle to teach about other concepts and ideas—forces, properties of matter, and fair testing—and more recently as a way to engage students in engineering practices and connecting to technology in the world around them.

The lesson I share here is just one of many lessons in a unit on magnetic forces and interactions. Students develop an understanding of the ways in which magnets interact with other objects and other magnets, and how these interactions can be used to solve problems. Many of the ideas about magnets that they build in this lesson are examined more deeply in subsequent lessons, and the questions that emerge during the activities often become jumping-off points for further exploration and learning.

Because of my extensive experience with the topic, the description that follows is an amalgamation of those lessons, and it represents many different students and teaching episodes as a single event. The instances I describe and the students I highlight are based on actual events, though I've taken creative license to make things flow in a single narrative.

See Table 5.1 (p. 94) for alignment of this lesson to the *Next Generation Science Standards* (*NGSS*; NGSS Lead States 2013).

## TABLE 5.1. NGSS Alignment

**Connecting to the NGSS—Standard 3-PS2-4: Motion and Stability: Forces and Interactions**

*www.nextgenscience.org/pe/3-ps2-4-motion-and-stability-forces-and-interactions*

- The chart below makes one set of connections between the instruction outlined in this chapter and the *NGSS*.
- The materials, lessons, and activities outlined are just one step toward reaching the performance expectation listed below.

**Performance Expectation 3-PS2-4.** Define a simple design problem that can be solved by applying scientific ideas about magnets.

| Dimensions | Classroom Connections |
|---|---|
| **Science and Engineering Practices** | |
| *Asking Questions and Defining Problems*<br><br>Define a simple design problem that can be solved through the development of an object, tool, process, or system and includes several criteria for success and constraints on materials, time, or cost. | Students "reverse engineer" products to identify the problem that is solved by using magnets. |
| *Planning and Carrying Out Investigations*<br><br>Make observations and/or measurements to produce data to serve as the basis for evidence for an explanation of a phenomenon or test a design solution. | Students devise a plan to test their ideas about magnets, including identifying appropriate materials, the data they should collect, and how to record their results. |
| **Disciplinary Core Ideas** | |
| *PS2.B: Types of Interactions*<br><br>Electric, and magnetic forces between a pair of objects do not require that the objects be in contact. The sizes of the forces in each situation depend on the properties of the objects and their distances apart and, for forces between two magnets, on their orientation relative to each other. | Students evaluate claims about magnetic interactions that relate to these core ideas (e.g., magnets must touch an object to attract it). |
| **Crosscutting Concepts** | |
| Interdependence of Science, Engineering, and Technology | Students identify the magnetic interactions within different products, how they are used to solve problems, and how these may represent improvements to existing items. |

## Engage Phase

I began this unit by telling my students, "At the end of this unit, we're going to be using magnets to solve some problems, but to do that, we need to understand more about what magnets can do. Do you have some ideas about what magnets can do?" Immediately, hands shot up and a few eager voices called out. "Wow! I can tell you have *a lot* of ideas," I said. "To make sure we give everybody a chance to share those ideas, I'm going to have you look over some statements about magnets and tell me whether you agree or disagree with them."

I used the formative assessment strategy of *agree/disagree* statements (Keeley 2008) to focus students on particular ideas about magnets. I provided these on a handout that included space to indicate "agree" or "disagree" or that "It depends …" followed by open space to write an explanation. The statements were as follows:

> **Teaching Tip:** I find that students are initially more comfortable debating claims that are presented to them than presenting their own claims for debate. Because these claims are also based on common misconceptions, they provide me with an opportunity to identify whether my students hold these misconceptions.

- Magnets attract metals.
- Magnets repel other magnets.
- Magnets can attract and repel through other materials.
- Magnets have to touch objects to attract them.

I asked the students to pair up and read the statements with a partner and explain in their own words what the sentence was claiming about magnets. This was important for students who may not have been familiar with these words in a scientific context. I watched students Jody and Maia as they read through the statements, Jody following along with her finger as Maia sounded out the words with her. I noticed students using hand gestures to explain to one another what *attract* or *repel* means. After the pairs were done, to check comprehension, I asked for volunteers to put the statements in their own words. We discussed where they might have heard those terms used elsewhere (e.g., "an attractive person," "bug repellent") and clarified how they refer to magnets moving toward and moving away from one another in this context.

**UDL Connection**

*Principle I: Representation*
*Guideline 2: Language and Symbols*
*Checkpoint 2.3: Support decoding of text, mathematical notation, and symbols*

I explained that each person now needed to decide whether they agreed or disagreed with the statement or thought that whether it was true depended on something else—and to write their ideas down. I noticed that while Brian was scribbling away his explanations, Luke had simply checked the box with his response. I decided that in the next part of the lesson, it would be important to ask him to communicate his ideas orally with peers. Providing multiple opportunities and ways to communicate ideas throughout the lesson (in writing, orally, or through pictures) is important so that all students can express their understanding.

I glanced at each page that the students were working on to make a quick survey of potential misconceptions they had, and to get a sense of the prior knowledge of the class. As students wrapped up the task, I called them back together and said, "OK, as I've walked around, I've noticed that some of you have *different* ideas about magnets—that's interesting! What I'd like you to do is share those ideas with each other and explain why you agree or disagree." I used a discussion strategy that provided students with an opportunity to share one-on-one with three other students, trading partners when I gave a signal. As the students rotated around the room to talk with different classmates, I eavesdropped on their conversations:

Maia: *I said I agreed that magnets can work through other things, because I have magnets on my refrigerator and they work through my paper.*

Luke: *Well, I sort of agree with you but I also sort of disagree, because it would depend on how much paper you had. Like, you couldn't stick this whole notebook of paper up on your refrigerator!*

Maia: *You are right, but I think if I had a strong enough magnet I could!*

**UDL Connection**

*Principle III: Engagement*
*Guideline 8: Sustaining Effort and Persistence*
*Checkpoint 8.3: Foster collaboration and community*

I could see where students were growing more comfortable in disagreeing with others' ideas—something we had practiced doing in socially appropriate ways. The students were acknowledging the other person's ideas, even if they didn't completely agree. To some extent, having them agree or disagree with claims I had provided them was more

**Teaching Tip:** Using think-pair-share involves *all* students in discussion, as opposed to a few, and ensures that everyone is getting to practice listening and speaking.

comfortable for them than presenting a claim of their own to the class for debate. It's much less risky to say whether you think someone else's idea is correct or incorrect than opening yourself up to the class to tell you whether your own ideas are correct. Because I was eliciting their prior knowledge in this phase, I didn't expect their ideas to be correct at this point—and so helping them feel safe in expressing their thinking was important to me.

**UDL Connection**

*Principle III: Engagement*
*Guideline 7: Recruiting Interest*
*Checkpoint 7.3: Minimize threats and distractions*

As I listened to students share, I noticed that their conversations elicited more ideas and experiences than they had initially considered—as well as some variables (e.g., magnet strength, thickness of the material, how you hold the magnet) that could affect whether the statement was accurate. The students had a lot of experience with magnets, but they had not made systematic observations of how they work in different situations. This was a cue to me that the students were ready to move on to the Explore phase.

## Explore Phase

It's tough bringing the class back together after lively conversations and unresolved differences in ideas—but those disagreements provided a great starting point for our explorations, as they created in students a need to know. "How would scientists resolve the differences in their ideas about magnets? Would they arm wrestle?" I joked. A few students picked up the humor and begin to mock arm wrestle. "OK, seriously, though—what would we do?" A chorus of voices chimed in with "Test it out! Do an experiment!" I told the students that that's exactly what they were going to do next.

I distributed a new handout to each pair of students, so that the original four statements were now divided up among the class. This allows each pair to confer with others investigating the same idea, and to contribute to the understanding of the whole class by sharing their expertise. Below each statement was the question "How could we find out?"

> **Teaching Tip:** Discussions can be boring when each group presents the same information over and over. By varying the claims that groups are assigned to test, students become more interested in learning about what their peers found.

"You and your partner need to come up with a plan. How you will test the idea?" I asked. "What kind of data do you need to collect? What materials will you use? When you have a plan, you may come to get approval and visit the materials center for supplies."

I found that the students had difficulty restraining the impulse to jump right into conducting their investigation—so it was important to support them by focusing on planning and its importance. Though it takes time, this planning is an essential part of the process—scientists don't just blindly begin testing through trial and error. The *NGSS* specifies that, as part of the practice of planning and carrying out investigations, students should think about the variables they might need to control and the number of trials they should consider, and they should evaluate the appropriate methods and the tools for collecting data.

**UDL Connection**
*Principle II: Action and Expression*
*Guideline 6: Executive Functions*
*Checkpoint 6.2: Support planning and strategy development*

My planning for the Explore phase required anticipating materials that would be useful to students in their tests. Before the students began, I reviewed the available materials so they would understand what they could use in their investigation. Students had access to several items; see the "Materials and Safety Notes" box, below.

---

### Materials and Safety Notes

**Materials**

Paper clips, nonlatex rubber bands, string, notecards, fabric, plastic cups, tape

Assorted magnets (weak and strong, in different shapes and sizes)

Assortment of metal objects, including some that are and are not attracted to a magnet (e.g., aluminum foil, brass key, steel bolt, copper penny, iron nail)

Scissors, measuring tape, rulers

Pan balance and mass set

**Safety Notes**

1. All involved must wear indirectly vented chemical splash goggles or safety glasses with side shields during all phases of these inquiry activities (setup, hands-on investigation, and take-down).
2. Direct supervision is required during all aspects of this activity to ensure that safety behaviors are followed and enforced.
3. Make sure that any items dropped on the floor or ground are picked up—a trip-and-fall hazard.
4. Use caution when working with sharp things such as metal objects, tools, and scissors. They can cut or impale skin.
5. Follow the teacher's instructions for returning materials after completing the activity.
6. Wash your hands with soap and water after completing the activity.

---

As students brought their plans to me, I looked over the plans and gave some a stamp of approval for them to pick up materials. In other cases, I posed questions to help the students refine their plans before I gave approval: "Would it matter what kind of material you use?" "How thick should the material be?" "What data will you record?" "What units will you use to record the distance?" "Will it matter which way you face the magnet?" I noticed that the more times we engaged in this process, the more detailed their plans were becoming and the quicker the approval process was completed.

As the students began their investigations, I circulated around the room asking questions to help them make sense of their observations and test their ideas further: "What are you noticing?" "How did that match what you thought before?" "Would it make a difference if you … ?" I noticed that two students who were testing

> **Teaching Tip:** Peer feedback pairs can also be used to help students review their plans before presenting them to the teacher.

the idea that magnets attract metals had sorted objects into two piles. They were puzzled because they noticed that metal objects were in both. One suggested that the things they thought were metal must not actually be metal (fitting the data to their initial idea that the statement was accurate). Fortunately, I kept samples of different metals that were labeled (iron, nickel, copper, zinc, steel, aluminum) in my supplies. I provided these to the students, which helped them identify that only iron and nickel are attracted to a magnet, as well as steel (which contains iron). As the pairs began to wrap up their work, I encouraged them to finish writing down their ideas to bring to small-group discussions.

## Explain Phase

I gave the signal for the class's attention and explained the next step—the pairs who had the same statements would meet to compare their ideas. In each corner of the room, I placed a sticky-notes chart paper with the original statement. I told the class, "Working together, I want you to decide if you think the statement is OK as written, or how it should be rewritten to be scientifically accurate." I explained that the groups would need to share their conclusions with the class and the evidence that supported their ideas. As I listened in on the conversations, I noticed that students were excited to find that their results agreed with other partners' results—making them more confident in their ideas. In cases where their results didn't agree, I was pleased to see students going back to the materials to retest their ideas.

After these discussions concluded, each group had a revision to the original statement they were assigned (which I had hoped for, given that they were intentionally worded in imprecise ways) written on a sentence strip. For example, the first group realized that magnets attract *some* metals, but not others. The second group also offered a revision to the original statement: "We think that magnets *can* repel other magnets but that it matters which way they are facing."

At that, a few students chimed in with "They have to be opposite poles ..." and "The positive and positive will repel." This signaled to me that students had potential misconceptions about the north and south poles of a magnet, which I knew would be addressed in a future learning cycle when we would test the poles of a magnet with an electroscope to determine whether they are, indeed, electrically charged (positive and negative).

"It sounds like we all agree that it matters which way they are facing," I summarized. "But we have some more investigating to do to understand why that matters."

**Teaching Tip:** Knowing when to address a misconception is hard. Rather than address this one when it came up in conversation, I made a mental note to revisit it in our future lesson. At this point, we had no evidence on which students could base an argument for defining a "pole" or differentiating whether poles are positive or negative, or north or south.

The third group provided evidence from their investigations with cups of water and stacks of notecards to support the claim that magnets can attract and repel through other materials, but they added that the thickness of the material was a factor, as well as how strong the magnet was. I made a mental note to connect back to this example when students develop a "fair test" to compare magnet strength in an upcoming lesson. These are both factors that can be investigated more systematically.

Finally, the fourth group demonstrated how bringing a magnet near an object could make them attract, disproving the claim that they had to touch the magnet to the object. They noted that the distance was different for different magnets—another promising idea to revisit as we devised methods to compare magnet strength. We posted all four of our new ideas on our science wall as a reference and read the sentences aloud together as a class.

"These are all really great discoveries you've made about how magnets work!" I encouraged the students. "And there are some people who have been able to use these same discoveries to design new tools and technology. We call those people engineers. In the next part of the lesson, we're going to take a look at the different ways that engineers have used magnets to solve everyday problems."

As a wrap-up for that day, I asked students to keep an eye out for magnets they encountered in their daily lives that were being used for different purposes.

Their homework was to find an example of a product that uses a magnet—something they used in their home, had seen on TV, or had talked about with a friend or family member. Their parents had already been given a heads-up about this task in advance in our class newsletter, along with suggestions if students got stuck finding an example.

**UDL Connection**
*Principle III: Engagement*
*Guideline 7: Recruiting Interest*
*Checkpoint 7.2: Optimize relevance, value, and authenticity*

## Extend Phase

We began the next session with a discussion of the different ways that students observed magnets being used in their daily lives. As the students shared their examples, I asked them to connect their examples to the four ideas about how magnets interact from our previous lesson. Maia brought up the magnets on her refrigerator again, connecting that to magnets being able to attract through other materials. She added, "But my magnet is not very strong because it could only hold up one of my drawings, not two."

I commended Maia for her use of evidence to support her ideas, and I noticed that the students who shared after her followed her example. Students brought up a variety of items including magnetic bumper stickers, cabinet doors, and lipstick cases. "Where did these items come from?" I asked. "The store!" one student called out, and a round of giggles ensued. "*Somebody* had to design them," I explained, "and those somebodies are engineers."

As a culminating project for the unit, the students would be designing something to solve a problem they identify—using what they have learned about magnets. This first lesson helped them begin to develop an understanding of what engineers do, and they would be formally introduced to the engineering design process later in the unit.

Because I know that the problem some magnetic products solve isn't always an obvious problem, we talked through a couple of examples together, recording our ideas in a chart on the board (see Table 5.2, p. 102). Julia explained that without the magnet, her mother's lipstick would come open inside her purse and get all over everything. "But her lipstick sticks to her keychain now!" she said, laughing. Luke offered up that the cabinet door magnet, like the lipstick lid, also keeps something closed.

Once we had identified the problems, we focused on the magnetic interactions that the engineers applied to solve the problem: "Is a magnet attracting or

| Product | Problem It Solves | How Magnets Are Interacting | Trade-Offs |
|---|---|---|---|
| Travel checkers | Checkers can fall and get lost in the car when you hit a bump, or they slide off their spaces when you go around a curve. | The board is a magnet and the pieces are metal (containing iron). They are attracted to the board. | The pieces are hard to move. If the magnet is not strong enough, they can still fall off. Other things can stick to the board as well. |
| Magnetic notepad | People lose their notepad. This lets you stick it to your refrigerator where you can easily find it when you need it. | The magnet is stuck to the notepad, and the magnet is attracted to the refrigerator door. | If the notepad is too heavy, it can slide down the refrigerator. Some refrigerator doors are not attracted to magnets. |
| Magnetic nametag | Regular nametags that use a pin can put holes in your clothes and can stick you unintentionally. | Two magnets attract each other through the fabric of your shirt. | The magnets have to be aligned to attract. It can be hard to pull them apart without dropping the one inside your shirt. These are more expensive than pins. |

TABLE 5.2. Sample Analysis Chart for Magnetic Products

repelling?" "Is it interacting with another magnet or another object?" "Is it acting through any other materials?" "Is the magnet making contact with something?"

It was also important that the students recognized the trade-offs or unintended consequences of design decisions that engineers make—I returned to Julia's example of her mom's lipstick case, and how other things stuck to it. "Do you think that was what the designer wanted to happen?" I asked. Students recognized that while a stronger magnet might be needed to hold the case shut, it might also be more likely to attract other objects.

We then reconvened for several groups to share their products and analysis before wrapping up the session for the day. I collected their handouts to review in preparation for the final phase of the lesson, so I could be sure to follow up with any potential difficulties that students were having.

## Evaluate Phase

Although I had students working together through much of the lesson, it was also important for me to understand what each student was taking away from the lesson individually. For this reason, I had the students work by themselves on the assessment activity as the final phase of the lesson. Because I had struggling readers, I offered several formats for accessing the information. I provided students with an array of examples of magnetic products (printed advertisements, actual products, or video clips showing the product); see Figure 5.2. I've gathered these over the years—and have received examples from fellow teachers and former students who knew about the activity as well!

---

**Figure 5.2.** Examples of Magnetic Products Featured in the Evaluate Phase Activity

### MAGNETIC PRODUCTS

➢ Magnetic nametags

➢ Magnetic travel games

➢ Magnetic poetry tiles

➢ Magnetic knife rack

➢ Magnetic key holder

➢ Magnetic makeup case

➢ Magnetic building tiles (toy)

➢ Magnetic hooks

➢ Magnetic notepads and calendars

➢ Magnetic fish-tank scrubber

➢ Magnet-tipped screwdriver

➢ Magnetic wristband to hold nails, screws, and washers

➢ Magnetic earrings or necklace

➢ Magnetic nail polish

➢ Magnetic eyelashes

➢ Magnetic eraser (for classroom whiteboard)

➢ Magnetic curtain rod

➢ Magnetic bumper stickers and signs

---

**UDL Connection**

*Principle I: Representation*
*Guideline 2: Language and Symbols*
*Checkpoint 2.5: Illustrate through multiple media*

---

I told the students, "I'd like you to do what we just did—choose an item. What problem does it solve? What magnetic interactions did the engineer use in the design of the product to solve that problem? Can you think of any potential trade-offs the engineer had to consider in his or her design? Are there any unintended consequences that might occur?"

I allowed the students to take time choosing a product that was of interest, and I explained that they also had a choice in how they communicated the answers to the questions—they would fill in the handout by writing their ideas, and using this information they could create a poster, video, or podcast.

**UDL Connection**
*Principle III: Engagement*
*Guideline 7: Recruiting Interest*
*Checkpoint 7.1: Optimize individual choice and autonomy*

Unlike many of the advertisements, the students' work should emphasize the science concepts in the design of the product, as well as the problem it solves. The chart we had made earlier provided a useful reference point for the kind of information that students should include in their work. This particular assessment focused on both the science and engineering practices and the disciplinary core idea for the performance expectation I had selected.

> **Teaching Tip:** We often end up using the daily instructional time for writing to edit and revise students' work on this project.

The students shared their work with their classmates as a final step in this process—and we discussed the variety of ways in which magnets are used. Students realized that many products represent improvements made to existing products (magnets + notepad = magnetic notepad). I encouraged them to think about products they knew of that could be improved with magnets, or other problems that magnets might help solve, as we continued our unit.

## Unpacking UDL: Barriers and Solutions

While reading through this lesson can help give you a sense of what happened in the classroom, I want to emphasize the *planning* that occurred prior to implementation. Below, I unpack additional ways in which I identified barriers that the curriculum and activities I chose to do might pose for particular students, and how I implemented some solutions. I didn't have to come up with entirely new activities or reduce expectations, but I was able to think about enhancing the activities to reduce those barriers and make the learning more accessible for my students. Table 5.3 summarizes the connections between the UDL framework and the overall design of the lesson. Following that are two examples of how UDL was applied to meet the needs of specific learners.

| TABLE 5.3. UDL Connections | |
|---|---|
| **Connecting to the Principles of Universal Design for Learning** | |
| **Principle I: Representation** | |
| **Guideline 2: Language and Symbols** | |
| Checkpoint 2.3. Support decoding of text, mathematical notation, and symbols | Paired reading can support struggling readers when handouts or printed materials are used, as in the agree/disagree statements activity in the Engage phase. |
| Checkpoint 2.5. Illustrate through multiple media | In the Evaluate phase activity, magnetic products are shown in real life, in print advertisements, and in video for students to comprehend and access content. |
| **Principle II: Action and Expression** | |
| **Guideline 5: Expression and Communication** | |
| Checkpoint 5.1. Use multiple media for communication | For the final assessment, students were provided a choice in how they communicated their thinking to others (written, pictorial, or oral) about the product they analyzed, so that even reluctant writers could demonstrate what they had learned. |
| **Guideline 6: Executive Functions** | |
| Checkpoint 6.2. Support planning and strategy development | To help students who may overlook strategic planning, they are provided time for planning and feedback on their plans prior to testing their ideas about how magnets interact. The agree/disagree statements from the Engage phase activity provide guidance for what to investigate and explore. |
| **Principle III: Engagement** | |
| **Guideline 7: Recruiting Interest** | |
| Checkpoint 7.1. Optimize individual choice and autonomy | Students are offered a choice of product to analyze in terms of identifying how magnetic interactions are used to solve a problem, to increase the degree of connectedness to learning. |
| Checkpoint 7.2. Optimize relevance, value, and authenticity | As a homework assignment and to increase the authenticity of the Evaluate phase, students are encouraged to identify objects they use daily that involve magnets and to use those as a basis for identifying magnetic interactions that are used to solve a problem. |
| Checkpoint 7.3. Minimize threats and distractions | Agree/disagree statements are a low-risk alternative for eliciting all students' initial ideas, which may or may not be correct. |
| **Guideline 8: Sustaining Effort and Persistence** | |
| Checkpoint 8.3. Foster collaboration and community | Using pair sharing provides students with an opportunity to develop their own ideas to share with the class. |

### Learner Profile: Maia

"Maia" was a struggling reader—and was reading far below the level of her peers. She liked working with partners and felt comfortable sounding out words that she couldn't read when working one-on-one. The Engage phase of this lesson included a series of *agree/disagree* statements on a written handout—which posed a barrier for Maia. Allowing her to partner up to read the statements and discuss their meaning first helped her be successful. This solution also helped foster communication and collaboration among all students—maintaining their interest in the activity as they moved on to testing those ideas against evidence.

During the individual assessment in the Evaluate phase of the lesson, relying on a printed description of a product would have made the task difficult for Maia—even though she may have understood the science ideas. Therefore, providing multiple options (actual products, printed ads with pictures, or video clips) ensured that Maia could comprehend and access the information. Providing multiple examples also supported the engagement of all students by giving them a choice of product they found to be personally interesting.

### Learner Profile: Luke

"Luke" was a reluctant writer. While he participated in lessons during writing instruction, writing in other content areas was problematic. Integrating work on some written portions of the lesson during our writing time allowed for a greater focus on areas in which he struggled. However, working with Luke challenged me to consider the *purpose* of student writing in science. I constantly asked myself, "Why am I having my students write? Are the students going to use their writing as a tool or revisit it? Is writing really necessary to learn the content or complete the task successfully?"

In the Engage part of the lesson, students were asked to write their reasoning for agreeing or disagreeing with the various statements. My intent was to give all students the necessary time to process and formulate their own ideas before sharing with others. I knew that some students in my class processed more slowly, and their thinking could be shut down once others started sharing their responses aloud. The writing time was a tool for thinking, to help students be prepared to discuss their ideas with peers. Thus, I wasn't concerned that Luke didn't write his ideas, because he still had thinking time to gather his thoughts so he could talk about his ideas. In contrast, in the Explain phase, *rewriting* the original statements to align with the new evidence and recording his ideas was an important task for Luke, as this could be referred back to as a tool for analyzing products later in the lesson.

In the Evaluate phase, communicating understanding was important—but I recognized that this communication could take multiple forms (verbal, written, pictorial). Having options for communicating in different media not only reduced the writing barrier for Luke, but also provided choice to *all* students, supporting their interest and engagement. As an added bonus, these formats also allowed students to share their ideas with others outside of the classroom, including parents.

---

### Questions to Consider

➤ To what extent did the activities in this lesson align with the purpose and intent of each phase of the 5E Learning Cycle? Could you envision other activities that would be appropriate for each phase?

➤ Were you able to follow the sequence of activities and the ideas that students developed across all five phases? How did the storyline of the lesson progress? How does this storyline build toward the culminating engineering design project for the unit?

➤ In what ways was the teacher able to assess students' ideas during each phase of the lesson? How did this inform her instruction in this lesson? How did this inform her plans for the rest of the unit?

➤ In what ways did the solutions that the teacher identified meet the needs of the specific students spotlighted in this vignette? In what ways did they benefit all students? Could you think of other solutions you might use for your own learners?

---

## References

Keeley, P. 2008. *Science formative assessment: 75 practical strategies for linking assessment, instruction, and learning.* Thousand Oaks, CA: Corwin Press.

NGSS Lead States. 2013. *Next Generation Science Standards: For states, by states.* Washington, DC: National Academies Press. *www.nextgenscience.org/next-generation-science-standards.*

# CHAPTER 6

# An Attractive Idea

### *Tracy Hager*

This chapter describes a third-grade lesson that is part of a larger unit on forces focused on the phenomenon of magnetic levitation and figuring out how a magnetic levitation ("maglev") train works. This lesson examines forces (pushes and pulls) in terms of attraction and repulsion between magnets and also looks at how magnets interact with other materials.

While my students are familiar with magnets, they have not systematically examined them to develop a deep understanding of the forces they exert. These forces are unique in that they do not require objects to be in contact with one another. The students might also believe, incorrectly, that magnets attract all metals, generalizing from their everyday experiences. Thus, this first lesson allows students to test some of their existing ideas against evidence as well as to develop new ideas about magnets. The outline of the conceptual storyline of the lesson follows (see Figure 6.1, p. 110). Because our daily time for science was short (30 minutes or less), the lesson was implemented over multiple class sessions—with each step informing the next.

**Figure 6.1.** Conceptual Storyline of the Lesson

# How Do Magnets Interact With Each Other and Other Objects?

**Engage**
Friendly talk probe

**Key idea:** Some people believe that magnets attract foil, and others do not

**Linking question:** Are our ideas correct?

**Explore**
Testing magnetic interaction

**Key idea:** Magnets do not attract non-metals and some metals (foil); they attract and repel other magnets

**Linking question:** What kinds of objects do magnets attract and repel?

**Explain**
Classifying objects based on interactions

**Key idea:** Magnets attract and repel each other; objects that contain iron attract magnets; magnets do not interact with other materials

**Evaluate**
Role play and reflection

**Key idea:** Objects in different groups interact in different ways

**Linking question:** How can we demonstrate the forces that magnets exert?

**Extend**
Mystery boxes—students identify concealed objects based on their interaction with a magnet

**Key idea:** Interactions help us make inferences about materials

**Linking question:** How can we use these interactions to identify objects?

## LESSON VIGNETTE

Unlike many elementary teachers, I have spent my entire career in third grade. During that time, I've seen many changes in standards and curricula; however, magnets is one science topic that has appeared (and reappeared) in the units studied in third grade. I remember when I first taught this topic, the children had so much fun rotating from one activity to another during our "magnet carnival." The silly faces they created on Wooly Willy using a magnet and iron filings delighted the children. Students used a magnet to move a monkey through a maze, racing their partner. They investigated the poles of a magnet, magnet strength, and many other magnet concepts as they progressed through the unit. I marked this unit as a success because the children enjoyed it, and they were engaged in many hands-on experiences.

What a contrast to how I teach magnets today! Rather than an end unto itself, our study of magnets has become a means to learn something bigger about forces. Over the years, I learned to use the 5E Learning Cycle as a way to sequence activities (Hager 2006). However, working with the Quality Elementary Science Teaching program helped me take that a step further to connect those activities into a coherent conceptual storyline. Universal Design for Learning (UDL) has helped me identify potential barriers that the activities I have selected may pose for my students, and I have been able to intentionally plan solutions to reduce those barriers.

While engagement and enjoyment of science are still important, "success" in science is more than that. Instead of simply being excited about implementing fun activities like I was as a new teacher, I now delight in nurturing students' deeper understanding through purposefully sequenced activities that build from the students' prior knowledge and take into account their diverse needs and abilities. I am working toward shifting my instruction from "learning about" to "figuring out," as emphasized in our new standards.

See Table 6.1 (p. 112) for alignment of this lesson to the *Next Generation Science Standards* (*NGSS*; NGSS Lead States 2013).

| TABLE 6.1. *NGSS* Alignment | |
|---|---|
| **Connecting to the *NGSS*—Standard 3-PS2-3: Motion and Stability: Forces and Interactions** | |
| *www.nextgenscience.org/pe/3-ps2-3-motion-and-stability-forces-and-interactions* | |
| • The chart below makes one set of connections between the instruction outlined in this chapter and the *NGSS*.<br>• The materials, lessons, and activities outlined are just one step toward reaching the performance expectation listed below. | |
| **Performance Expectation 3-PS2-3.** Ask questions to determine cause and effect relationships of electric or magnetic interactions between two objects not in contact with each other. | |
| **Dimensions** | **Classroom Connections** |
| *Science and Engineering Practices* | |
| *Planning and Carrying Out Investigations*<br><br>Make observations and measurements to produce data to serve as the basis for evidence for an explanation of a phenomenon. | Students observe how magnets interact with each other and with other objects. These interactions are important for figuring out the overarching unit phenomenon. |
| *Disciplinary Core Ideas* | |
| *PS2.B: Types of Interactions*<br><br>Electric, and magnetic forces between a pair of objects do not require that the objects be in contact. The sizes of the forces in each situation depend on the properties of the objects and their distances apart and, for forces between two magnets, on their orientation relative to each other. | Students examine how the orientation of two magnets influences their interaction, and they explore how magnetic forces can be felt through other materials (box lid). |
| *Crosscutting Concepts* | |
| *Cause and Effect*<br><br>Cause and effect relationships are routinely identified, tested, and used to explain change | Students examine what happens when they bring objects and magnets close to one another. |

## Engage Phase

At the beginning of this learning cycle about magnetic interactions, I wanted to pique my students' interest in magnets and understand their prior knowledge. There are many common misconceptions that students hold about magnets (Beaty, n.d.), among them that magnets stick to all metals. I chose to use a "friendly talk probe" (Keeley 2008) to investigate my students' thinking about this idea. I tore a piece of aluminum foil from a roll and held up a round magnet and the foil to my class. I announced that two students were discussing magnets, and that I wanted the class to think about which student they thought was correct. I provided my students with a handout but also projected the friendly talk probe onto the interactive white board, along with graphics of aluminum foil and a magnet.

Using the probe as well as showing the foil and magnet were designed to make the task authentic and novel, as a way to help my learners who struggled to attend interested and engaged. Further, to support the struggling readers, I read the following out loud:

> Student 1: *I think this piece of aluminum foil will stick to a magnet.*
>
> Student 2: *I don't think this piece of aluminum foil will stick to a magnet.*

**UDL Connection**
*Principle I: Representation*
*Guideline 1: Perception*
*Checkpoint 1.3: Offer alternatives for visual information*

**UDL Connection**
*Principle III: Engagement*
*Guideline 7: Recruiting Interest*
*Checkpoint 7.2: Optimize relevance, value, and authenticity*

I then asked the students to write about which student they agreed with and why. As students began writing in their notebooks, I took the magnet and foil to my students Christy and Mary and encouraged them to touch the foil and then hold the magnet. "Let's just go ahead and put them together!" Christy eagerly stated. I explained that I was interested in seeing what she thought would happen *before* we actually tried it. I continued circulating around the room to ensure that all students, particularly those kids who often lost track of things, were able to get started. For my struggling writers, such as Christy and Jacob, I asked them to verbally rehearse their ideas to help them construct what they were going to write. As they finished, I had them bring their notebooks to the carpet.

**UDL Connection**
*Principle II: Action and Expression*
*Guideline 5: Expression and Communication*
*Checkpoint 5.3: Build fluencies with graduated levels of support for practice and performance*

---

**UDL Connection**
*Principle III: Engagement*
*Guideline 8: Sustaining Effort and Persistence*
*Checkpoint 8.4: Increase mastery-oriented feedback*

---

> **Teaching Tip:** There are always ways to improve your lessons in hindsight. Providing a sentence frame such as "I agree with student _____ because _____" would have been a useful scaffold for the writing task for students, by making it explicit that they must choose one student and also include details in their explanation.

When all students had recorded their ideas and explanations, we compared them. Christy decided that the magnet would not stick because the aluminum foil was both hard and soft. Another child believed the foil and magnet would not stick because the foil was too weak. Still others thought the magnet would stick to the aluminum foil because it was metal, while some disagreed and thought that aluminum foil was not metal, so it would not stick. See Figure 6.2 for a sample student response.

To me, the students' reasoning was just as important as their answers. I was intrigued that some students did not identify aluminum foil as metal—some based this on their prior experience that magnets don't stick to foil and the misconception that magnets stick to all metals, while others misclassified the aluminum foil because it was more bendable (malleable) than other metals. This helped me think about the materials I would need to use during our explorations—I had planned a variety of metals (some that would attract magnets and some that would not) as well as non-metals. Using metals that varied in their properties or form (such as bendable aluminum foil and steel paper clips) would also be helpful to confront students' ideas. I told the students that, like scientists, they would be testing their ideas against evidence in the next activity.

## Explore Phase

In my classroom, students regularly sit in groups of four. These partnerships are carefully planned to maximize student learning and cooperation and to minimize potential distractions. I try to balance gender, reading abilities, and special needs at each table with the goal of students supporting each other in carrying out the tasks.

**Figure 6.2.** Example of Student Response

Name_____     Date_____

*Learning Cycle #1*

**Engage:**
Two students are discussing magnets.

Student #1: I think this piece of aluminum foil will stick to a magnet.
Student #2: I don't think this piece of aluminum foil will stick to a magnet.

Who do you agree with and why? Explain.

*I agree with student #2 because aluminum foil isn't metal and magnets stick to metal.*

**UDL Connection**
*Principle III: Engagement*
*Guideline 7: Recruiting Interest*
*Checkpoint 7.3: Minimize threats and distractions*

For this exploration, I chose to structure the process that the groups would use and guide it as a class. Each student received a data sheet with the names and pictures of 10 items on it, to ensure that all students understood what the items were. Please see the "Materials and Safety Notes" box for a list of the 10 items.

## Materials and Safety Notes

**Materials**

| | | |
|---|---|---|
| Pennies | Aluminum foil | Paper squares |
| Iron nails | Plastic cubes | Cotton balls |
| Steel paper clips | Glass marbles | |
| Steel washers | Rubber erasers | |

**Safety Notes**

1. All involved must wear indirectly vented chemical splash goggles or safety glasses with side shields during all phases of these inquiry activities (setup, hands-on investigation, and take-down).

2. Direct supervision is required during all aspects of the activity to ensure that safety behaviors are followed and enforced.

3. Make sure that any items dropped on the floor or ground are picked up—a slip/trip fall hazard.

4. Use caution when working with sharp things such as metal objects, tools, and scissors. They can cut or impale skin.

5. Follow the teacher's instructions for returning materials after completing the activity.

6. Wash your hands with soap and water after completing the activity.

I projected a matching data sheet on our interactive whiteboard and asked, "Which of these materials does your group think will stick to a magnet?" I then provided one member of each team a bag with the corresponding 10 items so their group could physically examine them. The person who received the bag took out one item and passed it around the group so everyone could share their observations and ideas. Then everyone wrote their own idea (whether or not they agreed with other members of their group) on their data sheet.

**UDL Connection**

*Principle I: Representation*
*Guideline 2: Language and Symbols*
*Checkpoint 2.1: Clarify vocabulary and symbols*

On my signal, the first person passed the bag of objects to the next person, who took out the second object and followed the same process. The groups then passed the bags to me when they were finished. I chose this routine of taking turns to distribute the objects to minimize conflict within groups and to maximize our learning time. For students who had difficulty maintaining attention, this procedure kept the predicting process moving without delays so they would stay engaged.

**UDL Connection**

*Principle III: Engagement*
*Guideline 7: Recruiting Interest*
*Checkpoint 7.3: Minimize threats and distractions*

The students' expectations for each object reflected the variety of responses from the Engage phase. Many students who responded that magnets stick to metals predicted that the magnet would stick to the aluminum foil and the copper penny along with the nail, paper clip, and washer. I was excited to see their reaction the following day when we would test each item using the magnet and they would find that their observations did not match their expectations!

**Teaching Tip:** Having the objects needed for the Explore phase prepared and in baggies helped my students be successful in materials management as compared to gathering the objects on their own. It also allowed us to quickly clean up, minimizing distractions during discussion and enabling easy retrieval of materials the next day as needed.

As we continued our activity in the next session, I knew that students might encounter some challenges working together during the testing of the objects. A few students had been upset with teammates who "wouldn't share" or who did not follow the directions during our passing of the materials and recording ideas the previous day. I created a teamwork reflection tool that encompassed our school-wide "Positive Behavior Intervention Strategies" goals: to be peaceful, respectful, and responsible. Under each expectation, I listed specific behaviors that students would follow during our science exploration and provided a place for each team to stop to reflect on their progress collaborating by assigning a smiley face, straight face, or sad face for each expectation three times throughout our lesson (see Figure 6.3, p. 118).

**UDL Connection**
*Principle III: Engagement*
*Guideline 8: Sustaining Effort and Persistence*
*Checkpoint 8.3: Foster collaboration and community*

**UDL Connection**
*Principle III: Engagement*
*Guideline 9: Self-Regulation*
*Checkpoint 9.2: Facilitate personal coping skills and strategies*

**UDL Connection**
*Principle III: Engagement*
*Guideline 9: Self-Regulation*
*Checkpoint 9.3: Develop self-assessment and reflection*

I asked the students how it would help them to be honest each time we reflected, and how they could work to have all smiley faces by identifying ways to improve. Only one team had all smiley faces by the final evaluation round, which reflected their honest self-evaluation. All teams, however, showed improvement throughout the lesson.

I circulated around the room as students recorded the interactions they had observed between the magnet and the variety of objects. Most of the students who had predicted that the magnet would stick to copper and aluminum foil tried to quietly erase their previous ideas and change them to have the "right" answer. Because I wanted to create a scientific community that values the learning process and reasoning rather than having a "right" answer, I encouraged students not to erase predictions that differed from the results—and I told them that it was something important to share in the discussion afterward. We wrapped up this phase of the lesson as the students finished recording their observations.

**Figure 6.3.** Sample Group Reflection on Teamwork

We Can Work as a Team!

How did your team do on each of these expectations?

😊 = Yes! We did it!  😐 = Not quite  ☹ = No. We will keep trying.

We were **peaceful:**    ①    ②    ③

    We used a level 2 voice.    ___ ___ ___

We were **respectful:**

    We shared materials.    ___ ___ ___

    We used kind words.    ___ ___ ___

    We solved problems quickly.    ___ ___ ___

We were **responsible.**

    We wrote our answers.    ___ ___ ___
    We followed directions.    ___ ___ ___

## Explain Phase

Students were now ready to begin making sense of the data from the Explore phase and to explain their current understandings. They began again in small groups with their bags of materials. I asked the students to look at the data they had recorded the previous day and to create three groups of objects. They should sort the actual objects into piles on their table and label each group with a sticky note based on the interaction they had used to put those objects together. The students also recorded this in their notebooks (see Figure 6.4).

I walked around to observe the students' groupings and noticed that although they had almost identical groupings, they had labeled them differently. Where necessary, I prompted particular students to look back at the data tables to observe patterns and recall what they had done, or I encouraged a group to make sure that all members of the group got an opportunity to share their ideas—even Mary, who used an iPad to communicate with. Many teams used the labels *magnets*, *stick to magnets*, and *don't stick to magnets* (see the photo on p. 120). One team labeled its groups *magnets*, *metals*, and *nonsticking*—signaling to me that these students recognized that some metals were in this latter group and weren't sure how to handle that. Another group used the labels *magnets*, *metal materials*, and *nonmetallic material*.

**Figure 6.4.** Sample Student Notebook From the Explain Phase

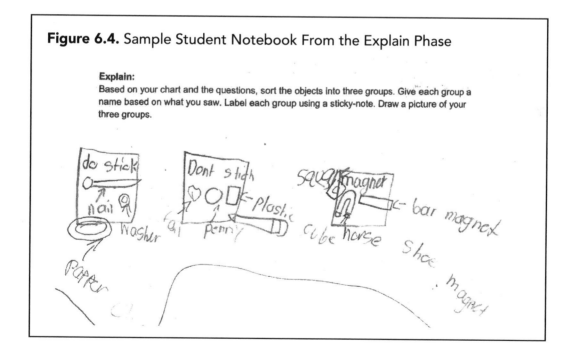

**Students sorted objects into three groups based on their interactions.**

**UDL Connection**

*Principle III: Engagement*
*Guideline 7: Recruiting Interest*
*Checkpoint 7.3: Minimize threats and distractions*

Then I gathered the teams on the carpet so we could share students' findings, clarify areas of inconsistency, and develop the appropriate scientific vocabulary to describe their ideas. For some of my students who needed help keeping focus, I sat them near me.

**Teaching Tip:** Students bring their notebooks with them to the carpet to use as a tool to engage in discussion, and to help them provide evidence to support their ideas.

On the interactive whiteboard, I displayed labeled pictures of the objects that the teams had tested. These graphics could be manipulated on the screen and placed into groups. I have found that the interactive whiteboard serves as an excellent tool for students who need extra support to keep engaged, and it provides a way for them to communicate their ideas to the whole-class discussion, particularly if they are nonverbal or have speech difficulties (such as with Mary and Christy).

The first group demonstrated what they did by moving the three magnets together. I circled that and asked what we could call this group. In unison the class exclaimed, "Magnets!" I asked the students to describe how the magnets interacted with each other. Louis mentioned that they stuck together. "Did they move toward each other?" I paraphrased. Louis nodded yes. "Let's show this action with our hands," I said. We put up our hands and brought them together to clap. "We learned from the beginning of our force unit that force is a push or pull. Would you say this is a push or pull force?"

Some of my students needed additional support to remember and transfer ideas from one learning situation to another, so I often used hand motions to help the students understand new vocabulary by associating each with a kinesthetic motion. I elaborated as I moved my hands together that the magnets *pull together*. I said, "Scientists have a word they use to describe two objects that pull together: attract." I labeled *attract = pull together* on our interactive whiteboard page next to the magnet group. "Let's show *attract* with our hands pulling together as we say the scientific vocabulary term," I instructed.

**UDL Connection**
*Principle II: Action and Expression*
*Guideline 5: Expression and Communication*
*Checkpoint 5.1: Use multiple media for communication*

**UDL Connection**
*Principle III: Engagement*
*Guideline 7: Recruiting Interest*
*Checkpoint 7.3: Minimize threats and distractions*

**UDL Connection**
*Principle I: Representation*
*Guideline 3: Comprehension*
*Checkpoint 3.4: Maximize transfer and generalization*

Next, I wondered aloud, "Was there any other kind of interaction between the magnets?" Earlier, I had seen Cormac pushing the magnets around on his desk during the Explore phase. As I caught his eye, he eagerly raised his hand to share what he had noticed. "I could push the bar magnet around on my desk with the round magnet," he said. "It moved away from me."

"Let's show this action with our hands," I encouraged. We started with our hands together and then moved them apart. "When two magnets push apart, scientists say they *repel*." We again repeated our new science term while showing the action. I wrote *repel* = *push apart* next to the magnet group as well.

> **Teaching Tip:** I often try to model good habits of inquiry. Wondering aloud illustrates to students how they should be thinking about their data and making sense of it.

To further clarify vocabulary, I connected these words to students' background knowledge and how they might use these words in everyday life. For example, many of my students had spent time in the outdoors in Missouri, and they had encountered mosquitos and other insects that bite. I shared, "When I go camping, I spray insect *repellent* on my arms and legs. Why? I want those mosquitos to go away from me!" I made the same motion as I had for *repel* with magnets, much to the students' delight. "When I go camping, I also notice that when I have my lantern on, the bugs all fly to it—they are *attracted* to the light," I said.

**UDL Connection**
*Principle I: Representation*
*Guideline 3: Comprehension*
*Checkpoint 3.1: Activate or supply background knowledge*

Next, we proceeded to the objects that students had found stuck to the magnet. Student volunteers came up and moved the graphics of the nail, washer, and paper clip into this group. All students indicated agreement with their grouping. "What types of interaction did you observe between the magnets and these objects?" I probed. Students responded that they moved toward the magnet and stuck to it. "What scientific term could we use to describe that?" I asked. We put up our hands and brought them together to demonstrate and repeat "attract = pull together." I labeled this force on our interactive whiteboard page. "Scientists have a word for these types of objects that are attracted to magnets—*ferromagnetic*." I underlined the *Fe* at the beginning of *ferromagnetic* and shared that *Fe* in science stands for iron. Two of my students called out excitedly to share that *Au* stands for gold. Although many students did not have background knowledge of the periodic table, this provided a link to the spelling of the word and also its meaning. "These objects all contain iron," I explained.

We moved on to the remaining objects that formed the third group—materials that did not stick to the magnet. Students were a bit puzzled as to how to describe these objects, because they neither attracted nor repelled the magnet. Because we had used hand motions for attract and repel, I suggested holding up our arms and crossing them to show there was *no interaction* between each object and the magnet. I introduced the term *nonferromagnetic* and wrote it next to this group on the interactive whiteboard. I activated students' prior knowledge that the prefix *non* means *not*, and I explained that this big, long word simply means that it does not contain iron. I then asked them to brainstorm other words they knew that had this prefix. Examples shared included *nonfat* ("My mom likes nonfat yogurt!") and *nonsense* ("Like when you do not make sense!").

**UDL Connection**
*Principle I: Representation*
*Guideline 3: Comprehension*
*Checkpoint 3.1: Activate or supply background knowledge*

Next, I provided each student with a set of color-coded response cards with words and pictures to represent the interactions of each group: objects that are magnets, objects that are ferromagnetic, and objects that are nonferromagnetic. The pink magnet card (see Figure 6.5) included words and pictures to illustrate *attract* and *repel*. The yellow ferromagnetic card (Figure 6.6, p. 124) included words and pictures to illustrate *attract*. The green nonferromagnetic card (Figure 6.7, p. 124) included pictures and words to express *no interaction*. I told the students that I would pull out an object and then count to 3. On 3, they should hold up the card for the group to which they felt the object belonged.

**Figure 6.5.** Magnet Vocabulary Card (Pink)

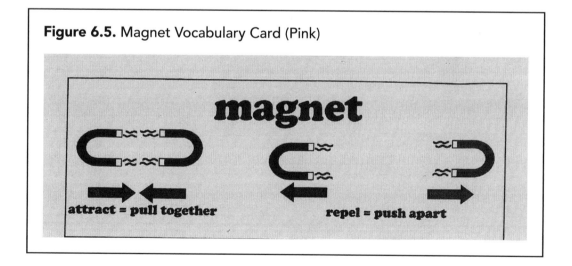

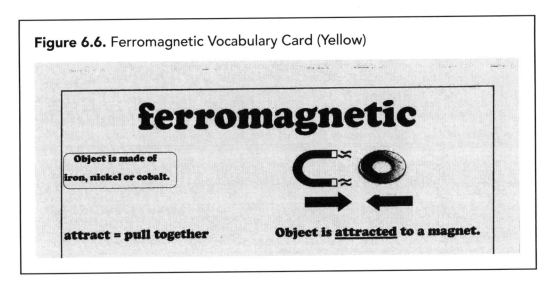

**Figure 6.6.** Ferromagnetic Vocabulary Card (Yellow)

**ferromagnetic**

Object is made of iron, nickel or cobalt.

attract = pull together

Object is <u>attracted</u> to a magnet.

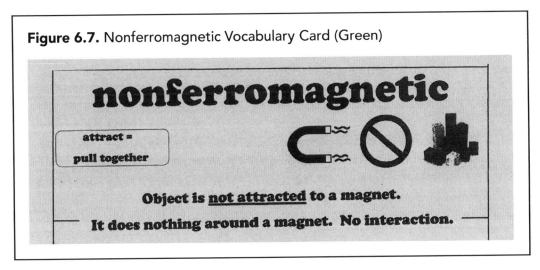

**Figure 6.7.** Nonferromagnetic Vocabulary Card (Green)

**nonferromagnetic**

attract = pull together

Object is <u>not attracted</u> to a magnet.

It does nothing around a magnet. No interaction.

As I held up each item and observed the cards that students held up in response, I was able to quickly check with a visual scan whether the students could correctly classify each item. Based on their earlier explorations, they could. However, I decided to pull out a *new* item (a metal fork) to challenge students' thinking further. The sea of cards was a mixture of yellow and green, and some students held up no card. I paused for students to think-pair-share. "I don't know what it's made of," I heard one begin. "Well, it's silverware so that's not iron," her partner responded. After the students had time to gather their thoughts, I had several of them share with the class.

Student discussion centered on the fact that metals could be *either* ferromagnetic *or* nonferromagnetic, as there were metal objects in both categories. I reminded students that ferromagnetic objects contain iron and asked whether they could tell if something contained iron just by looking at it. Some confidently claimed they could, while others suggested we needed to test it out. "What would it mean if a magnet is *not* attracted to this fork?" I asked. "What would it mean if it *is*?" I instructed the students to think-pair-share again to allow everyone an opportunity to express their answer, as opposed to just calling on one student. Then, I asked for volunteers to share what their partners had thought.

We reached a consensus as a class that if the magnet attracted the fork, it would mean that the fork contained iron. You can imagine some students' surprise when I picked up the magnet and stuck it to the fork! I explained that though we call forks "silverware," they can actually be made of many different things. "Like plastic!" one student called out. I then told the students to hold up the card for the group the fork belonged to, and I saw a sea of yellow. (See the photo below.) Picking up on the previous student's example, I asked what card they would choose if the fork were instead *plastic*. This time, the colors shifted to a sea of green. Then we wrapped up the activity for the day.

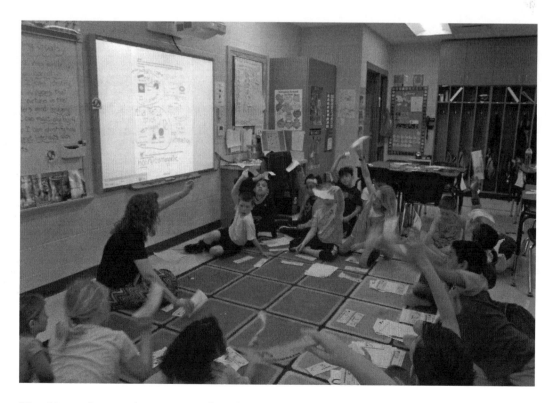

**Mrs. Hager has students respond with color-coded cards.**

## Extend Phase

While in the previous phase students *created* the groups based on how objects interacted, in this phase the students used observations of interactions to infer to which group objects belong. I envisioned a whole-class role-play activity in which each child would be one of the three kinds of materials: magnetic, ferromagnetic, or nonferromagnetic. I did this as a way to help students who needed additional support to process ideas for understanding and recall. However, I could foresee my students who have difficulty cooperating with others becoming very frustrated. Another barrier would be that students with learning disabilities and others who have difficulty understanding the scientific terminology would not be able to successfully participate in the role-play. Furthermore, students who struggle to maintain focus might also have difficulty staying on task. To reduce these barriers, I decided to restructure the activity so that the students would interact in small groups.

**UDL Connection**

*Principle I: Representation*
*Guideline 3: Comprehension*
*Checkpoint 3.3: Guide information processing and visualization*

One table of four students performed a role-play at a time. When the team came up to me, we huddled around a basket of objects. I assigned two students in each team to be magnets. I gave those students each a magnet and a card showing *attract* and *repel*. These students would need to demonstrate both of these interactions in the role-play. (See the photo on the next page.) For the first group, we chose two objects—a washer (ferromagnetic material) and a plastic cube (nonferromagnetic material). I brainstormed some ideas with the two students who were magnets for how to demonstrate *attract* and *repel*. The student who was the washer chose one magnet to stick to throughout the entire role-play. The student who was the cube stayed in the same place for the entire role-play with no interaction with the other group member.

**UDL Connection**

*Principle III: Engagement*
*Guideline 7: Recruiting Interest*
*Checkpoint 7.3: Minimize threats and distractions*

I created a recording sheet (see Figure 6.8, p. 128) on which I included the information from the color-coded response cards at the top. I provided a three-column table in which students were to write the participant's name, record interactions that they saw, and circle which material they decided it was based

**Students role-play interactions (attracting and repelling) between unidentified objects.**

on the interactions. For my students with attention concerns, this focused their attention on only four materials to analyze at a time. Rather than being distracted by 25 students milling about asking questions, they could focus on one team at a time. I removed extra distractions to enable students to focus on a small amount of essential information. When their team performed, they needed to accurately demonstrate the interactions. As an audience member, students were responsible for watching the role-play, recording the interactions, and circling which material each child was demonstrating. This provided individual accountability.

**UDL Connection**

*Principle II: Action and Expression*
*Guideline 6: Executive Functions*
*Checkpoint 6.3: Facilitate managing information and resources*

Each role-play started with students standing on one of the four corners of our classroom rug. Then I would say "Begin," and the students would move to show *attract*, *repel*, or *no interaction*. After a few minutes, we would stop the role-play and they would write one participant's name on their recording sheet.

**Figure 6.8.** Student Recording Sheet Used in the Extend Phase

## Guess That Material

name:

Learning Goal: I can identify the material by watching a role play showing interactions.

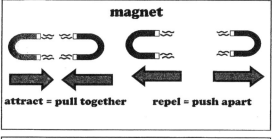

**magnet**

attract = pull together     repel = push apart

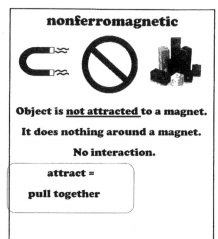

**nonferromagnetic**

Object is <u>not attracted</u> to a magnet.

It does nothing around a magnet.

No interaction.

attract =

pull together

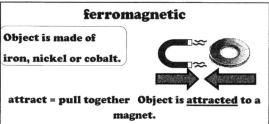

**ferromagnetic**

Object is made of iron, nickel or cobalt.

attract = pull together   Object is <u>attracted</u> to a magnet.

| Name | Interactions I See:<br>(attract, repel, nothing) | What Material is it?<br>Circle your answer. |
|---|---|---|
| Mrs Hager | attract<br>repel | magnet<br>ferromagnetic<br>nonferromagnetic |
| | nothing | magnet<br>ferromagnetic<br>nonferromagnetic |
| | attract<br>repel | magnet<br>ferromagnetic<br>nonferromagnetic |

That participant showed us again his or her interactions with the other materials. As a class, students used their hand motions to signal the direction of the force. They recorded it on the sheet. Next, each student held up a color-coded response card to show which kind of material was being acted out. From a quick scan of the room, I easily assessed student understanding by seeing if all students held up the same color or if we had a sea of different colors, which would indicate confusion and a need for additional clarification. This repetition of activities along with scaffolds for completing the task provided excellent opportunities for helping the students who often needed repetition and feedback to solidify and communicate what had been learned.

**UDL Connection**
*Principle II: Action and Expression*
*Guideline 5: Expression and Communication*
*Checkpoint 5.3: Build fluencies with graduated levels of support for practice and performance*

Following all the role-plays, the students independently drew a model of the interactions in their magnet packet. They needed to show how they interacted with another student in their team during their group's role-play, draw arrows to show the direction(s) of the force, and label each material as ferromagnetic, nonferromagnetic, or a magnet. The model demands that students combine all four roles and show the direction of the forces using arrows by drawing the model. They also needed to label each material; this included synthesizing the information of three types of materials and all four roles from the role-play.

During the role-play, students expressed their understanding of the interactions of their material and others via moving or not moving. As an observer, students demonstrated their understanding of each interaction by writing the information on the recording sheet. As shown in Figure 6.9, Addison accurately

**Figure 6.9.** Addison's Drawing Explaining the Interactions From the Role-Play

labeled each student as *magnet*, *ferromagnetic*, or *nonferromagnetic*. In showing the interactions between materials, she used the correct symbol for no interaction with the nonferromagnetic object. She also showed arrows indicating attraction between the magnet and the ferromagnetic object. She drew the second magnet from the role-play on the opposite side of the ferromagnetic object and showed repelling, rather than drawing the second magnet near the first magnet. After looking over her model, I realized I needed to investigate further to determine whether Addison was confused about the interactions or perhaps just needed to revise where she drew her second magnet on her model.

While most students who were assigned to be magnets during the team role-play demonstrated both attracting and repelling, only one student in the class showed both interactions on her model (see Figure 6.10). Although Cecilia did not label the participants with *magnet*, *ferromagnetic*, and *nonferromagnetic*, because I assessed those models immediately following the role-plays, I remembered which student had each role. Cecilia correctly included both *attract* and *repel*, using her arrows to show the direction of the force, or no interaction, for each role. Even though the students accurately showed the interactions for their material during the role-play, most students did not include both *attract* and *repel* on their model. Based on this, I realized that the students needed additional opportunities to clarify the interactions and practice new vocabulary.

**Figure 6.10.** Cecilia's Drawing Explaining the Interactions From the Role-Play

## Evaluate Phase

The purpose of the Evaluate phase is to assess student learning, but it's important that students are prepared to demonstrate their learning. Based on students' models from the role-play activity, I thought they needed some additional experiences with actual materials to help them clarify and express their understanding. I chose to focus the students' attention on the interactions between a mystery object and a magnet. Using pizza boxes, I created mystery boxes with six objects taped inside the lid. On the outside of the lid, above each of these objects, were six circles drawn and labeled with a letter, *A* through *F*. (See the photo below.) Students examined how a magnet interacted with each of these test spots. By observing the cause and effect (bringing a magnet near and the resulting interaction), they could make inferences about the object underneath. Based on the interactions they felt (attract/repel, attract only, or no interaction), they circled which kind of material they thought was under each spot. By increasing the number of mystery boxes available and decreasing the number of students who needed to share supplies, the engagement and cooperation levels increased.

**Students observe interactions between a magnet and items concealed in a mystery box.**

To support my students with learning disabilities and others who had difficulty understanding and retaining concepts, I copied the color-coded response vocabulary cards to the top of the recording sheet to serve as a prompt (see Figure 6.11). As a solution for the students with attention concerns and for those who had difficulty cooperating with others, I chose to have students work in specific partnerships. I modeled how one partner would first hold the magnet in different positions above a test spot, noting the interactions, and then hand the magnet to his or her partner to follow the same procedure.

**UDL Connection**
*Principle II: Action and Expression*
*Guideline 6: Executive Functions*
*Checkpoint 6.3: Facilitate managing information and resources*

**UDL Connection**
*Principle III: Engagement*
*Guideline 7: Recruiting Interest*
*Checkpoint 7.3: Minimize threats and distractions*

**UDL Connection**
*Principle III: Engagement*
*Guideline 8: Sustaining Effort and Persistence*
*Checkpoint 8.3: Foster collaboration and community*

The students were not allowed to open their mystery box until they had met with me. Together we reviewed the interactions they felt, and which type of material they thought was underneath each test spot. We then opened the box to see if they were correct. Overall, students were successful in identifying the kinds of materials found under each test spot in the mystery boxes; however, some students had difficulty feeling the attraction between their magnet and the small washer. This was difficult to distinguish because it was a very small pull. Once the box was opened, they were able to visually discern the washer moving toward their magnet.

> **Teaching Tip:** It's good to use larger objects that will enable students to easily identify them through tactile experiences with the mystery boxes.

As a final, individual evaluation, students responded to a friendly talk probe (Keeley 2008) written in a comic-strip format. A girl and a boy are shown. The girl is holding a magnet in her hand and another object that's attracted to it. The girl states, "I think I am holding a ferromagnetic object because my magnet

**Figure 6.11.** Sample Student Notebook Page From the Mystery Box Activity

## Mystery Box Objects

name: ‗

Learning Goal: I can identify a material by observing its interactions with a magnet.

**magnet**

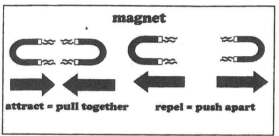

attract = pull together    repel = push apart

**nonferromagnetic**

Object is <u>not attracted</u> to a magnet.

It does nothing around a magnet.

No interaction.

attract =

pull together

**ferromagnetic**

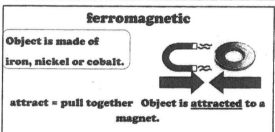

Object is made of
iron, nickel or cobalt.

attract = pull together    Object is <u>attracted</u> to a
magnet.

| Mystery Box Object | Interactions I See: (attract, repel, nothing) → ←← → | What Material is it? Circle your answer. |
|---|---|---|
| A | Attract | magnet<br>~~ferromagnetic~~ (circled)<br>nonferromagnetic |
| B | repel attract | magnet (circled)<br>ferromagnetic<br>nonferromagnetic |
| C | attract | magnet<br>ferromagnetic (circled)<br>nonferromagnetic |

is attracted to it." The boy counters by saying, "I think you're holding a magnet because your magnet was attracted to it."

The students were provided a space to insert themselves and their own comments in response to the following: "These two students disagree. What would you say to settle this disagreement?" I provided support by reading the probe out loud to students who had difficulty reading.

**UDL Connection**
*Principle I: Representation*
*Guideline 1: Perception*
*Checkpoint 1.3: Offer alternatives for visual information*

All students demonstrated some level of understanding of possible interactions in their responses. Four students wrote that the girl was correct because the object attracted. While this is one possibility, they did not take into account explicitly that magnets should both attract *and* repel. Two students replied that both were correct because a ferromagnetic object and a magnet could both stick to a magnet. Although no one suggested that the boy and girl try to see if the object could be repelled, some students did write that in order for the object to be a magnet, it must both attract and repel. Their responses helped me identify a need to scaffold their coordination of claims and evidence in upcoming lessons.

## Unpacking UDL: Barriers and Solutions

While reading through this lesson can help give you a sense of what happened in the classroom, it should be emphasized that the planning occurred before implementation, specifically with regards to applying the principles of UDL. Although some of the students had disabilities that are not common in typical classrooms (agenesis of the corpus callosum and seizure disorder), the barriers they faced were not unique: difficulty remembering concepts, issues with social relationships and getting along with others, trouble maintaining engagement in the content, and the need for support to know how to record information. The solutions for these barriers can be helpful for a large percentage of learners.

Table 6.2 summarizes the UDL principles, guidelines, and checkpoints that were applied when the activities were designed for each phase of the 5E Learning Cycle lesson to meet the *general* needs of the learners in the classroom. The sections that follow detail three examples of how barriers were identified and solutions were strategized to meet the *specific* needs of some of the learners in the classroom.

| TABLE 6.2. UDL Connections |
|---|
| **Connecting to the Principles of Universal Design for Learning** |

| Principle I. Representation | |
|---|---|
| **Guideline 1: Perception** | |
| Checkpoint 1.3. Offer alternatives for visual information | During the Engage and Evaluate phases, the prompts were read aloud to ensure that struggling readers could access information. |
| **Guideline 2: Language and Symbols** | |
| Checkpoint 2.1. Clarify vocabulary and symbols | During the Explore phase, vocabulary was presented with both words and pictures to ensure that all students could understand what the items were. |
| **Guideline 3: Comprehension** | |
| Checkpoint 3.1. Activate or supply background knowledge | Everyday examples during the Explain phase were used to develop understanding of key vocabulary. |
| Checkpoint 3.3. Guide information processing and visualization | During the Extend phase, a role-play was used as a way to further develop understanding of critical ideas, particularly for the students who took longer to process ideas. |
| Checkpoint 3.4. Maximize transfer and generalization | Some students struggled to recall and transfer ideas. To support those students, key concepts were connected with hand actions during the Explain phase. |
| **Principle II. Action and Expression** | |
| **Guideline 5: Expression and Communication** | |
| Checkpoint 5.1. Use multiple media for communication | An interactive whiteboard displaying graphics of the objects used during the Explore phase served to communicate ideas during the Explain phase to support students who struggled to communicate orally. |
| Checkpoint 5.3. Build fluencies with graduated levels of support for practice and performance | Struggling writers were given an opportunity to verbally rehearse their ideas prior to writing a response during the Engage phase. |
| **Guideline 6: Executive Functions** | |
| Checkpoint 6.3. Facilitate managing information and resources | In the Extend and Evaluate phases, students were provided a recording sheet to help them record their findings, particularly for those who had difficulty managing information. |

*(continued)*

| TABLE 6.2. UDL Connections (*continued*) | |
|---|---|
| **Principle III. Engagement** | |
| **Guideline 7: Recruiting Interest** | |
| Checkpoint 7.2. Optimize relevance, value, and authenticity | Real objects and an authentic task such as the ones used during the Engage phase helped students attend to the task and stay engaged. |
| Checkpoint 7.3. Minimize threats and distractions | Strategic group or partner arrangements, such as those used during the Explore and Evaluate phases, provided a way to create a supportive classroom climate for carrying out tasks. <br><br> Clear expectations and routines were provided during hands-on tasks in the Explore phase and the role-play in the Extend phase to ensure that students who experienced difficulty working with others would participate. <br><br> During the Explain phase, students were given prompts to ensure that they all had an opportunity to participate in group discussion. |
| **Guideline 8. Sustaining Effort and Persistence** | |
| Checkpoint 8.3. Foster collaboration and community | Students were placed in partnerships to complete activities together during the Explore and Evaluate phases to foster and develop collaboration skills. |
| Checkpoint 8.4. Increase mastery-oriented feedback | During the Extend phase, color cards were used to quickly evaluate student responses to provide for feedback. The role-play was carried out multiple times to provide repetition to help students solidify ideas and communicate what they knew. |
| **Guideline 9. Self-Regulation** | |
| Checkpoint 9.2. Facilitate personal coping skills and strategies | During the Explore phase, students were provided a checklist of behaviors to use to complete a group activity. This served as a reminder and a model for the students who experienced difficulty working with others. |
| Checkpoint 9.3. Develop self-assessment and reflection | During and after a science exploration in the Explore phase, students completed a self-reflection check to identify ways to improve group collaboration. |

### Learner Profile: Mary

"Mary" was an affectionate young lady who always had a smile. She was born with a brain disorder that adversely affected her academic achievement, communication skills, and executive functioning. The speech center of her brain was also affected, so she could not speak but instead used an iPad with the Touch Chat app to communicate her thoughts. She had an individual education plan (IEP) for "other health impairment" and received services for reading, writing, speech, and math. Though she received paraprofessional support in her special education classes, she did not have a para in her general education third-grade classroom.

Her primary challenges included difficulty communicating orally and in writing, organizing her ideas and recording them accurately, and cooperating with others. Mary was also easily distracted by other things happening around her and got very tired by the end of the day, finding it difficult to concentrate on the task at hand.

Throughout the learning cycle, I implemented a variety of solutions to support Mary in learning the concepts and communicating her understanding:

- In the Engage phase, I provided the actual aluminum foil and magnet for Mary to touch as she decided whether she thought the magnet would stick to the foil. This increased her engagement and also helped her know what items the probe was referring to.

- The Explain phase posed many challenges for Mary. She needed to cooperate with her group to sort the items. Because she primarily communicated with her iPad, it was easy for group members to not solicit her ideas and to plow on without her input. I prompted the groups to make sure they were gathering the opinions of every group member. Following along with the explanations of the groups and the direction of the forces for magnetic, ferromagnetic, and nonferromagnetic objects required maintaining mental focus. I seated Mary right next to me and regularly included her in the Explain phase by having her move an item into a group on the interactive whiteboard and share her ideas using her iPad. I did this to encourage her contribution to and engagement with the class discussion.

- In the Extend phase, participating accurately in the "Guess That Material" role-play required Mary to remember the interactions between magnets and other materials and to show it through movement during her team's role-play. I specifically chose to assign her the role of a nonferromagnetic object because that type of material has no interaction with a magnet. She could just stay in the same place on the rug where she started, and it reduced the level of complexity of the task for her to be successful.

- The Evaluate phase was structured in partnerships for the entire class to increase cooperation and to give students more hands-on opportunities as compared to groups of four. To support Mary, I specifically chose a partner who encouraged her to share her ideas and who was eager to work with her as opposed to doing the task for her.

### *Learner Profile: Christy*

"Christy" was an enthusiastic girl who loved school. She had a seizure disorder, and the medication she took to prevent seizures caused short-term memory loss, irritability, and sleepiness. Her IEP included goals for developing executive functioning skills. She received services for speech, reading, writing, and math. Christy did not receive paraprofessional support.

Christy's primary challenges included difficulty recalling information, mood swings, and difficulty cooperating with others. These factors affected not only her learning, but also her relationships with other children in the classroom. She frequently put her materials away in her desk if there was a break in activity (even in group activities), whether or not we had finished a task. Instead of picking up where she had left off, she drew pictures or read a book. Similarly, when frustrated with difficult tasks, she gave up and switched to another task. During some group activities, she struggled to participate. When she perceived that her input was not valued or if she was corrected for inappropriate behavior (such as not sharing materials), she responded by screaming at her group members.

Throughout the learning cycle, I implemented a variety of solutions to support Christy in learning the concepts and actively participating in the learning activities with her group:

- In the Engage phase, I provided the actual aluminum foil and magnet for Christy to touch as she decided whether she thought the magnet would stick to the foil. This increased her engagement and also helped her know what items the probe was referring to. I also had her verbally tell me her answer to rehearse it before she wrote it in her science student packet. This allowed her to construct her explanation orally when I could support her in fully explaining her ideas before she wrote her thinking, which added another level of complexity to the task. Then I pointed to where she should write her answer.

- During the Explore phase, when Christy was testing the items with a magnet, I regularly stopped by her desk to point out where she should record her results. This helped her keep track of where to write the responses in the data table. I also intervened in her group several times to encourage her to follow the group procedures of passing the objects and magnet around to each team member. She was so excited to test the actual objects that she was reluctant to pass them around to the rest of the children in her group. I provided reminders about the procedures for testing whether or not each object would stick to a magnet.

- The Explain phase posed many challenges for Christy. She needed to cooperate with her group to sort the items and often did not like following directions or sharing the materials. I prompted the groups to make sure they were gathering the opinions of each group member. Christy struggled with higher-level thinking, so although she enjoyed talking and interacting with the other students at her table, she had difficulty identifying patterns to group objects together that had similar characteristics. She would rather manipulate and play with the objects than think about ways to group them into categories. Because we had a day between our Explore and Explain phases, she needed lots of questions and scaffolding from me and to look back at her recording sheet at what happened when she tested the objects, due to her short-term memory loss.

- When all the groups convened on the carpet to share their results, I seated Christy right next to me and regularly included her in the Explain phase by having her move an item into a group on the interactive whiteboard. Following along with the explanations of the groups and the direction of the forces for magnetic, ferromagnetic, and nonferromagnetic objects required maintaining mental focus. Seating Christy next to me helped keep her engaged and following along with the lesson. She thrived on attention and praise from the teacher and was very motivated by it. Often during a lesson if I incorporated her ideas into an explanation indicating that she was correct, she would throw her arms around me and give me a big hug. Just sitting next to me she would often grab me in a spontaneous hug of joy. Seating her next to me minimized threats and distractions and kept her engaged in the lesson.

- In the Extend phase, participating accurately in the "Guess That Material" role-play required Christy to remember the interactions between magnets and other materials and to show it during her team's role-play. This was a challenge due to her short-term memory loss. I specifically chose to assign her the role of a nonferromagnetic object because that type of material has no interaction with a magnet. I provided additional support by verbally rehearsing with her and planning her role of staying in the same spot on the rug when I called "Begin." This rehearsal helped her remember her role and supported her in being successful during the role-play.

- The Evaluate phase was structured in partnerships for the entire class to increase cooperation and to give students more hands-on opportunities as compared to groups of four. To support Christy, I specifically chose a partner who encouraged her to share her ideas and who was happy to work with her as opposed to doing the task for her. I regularly stopped by the pair to assist Christy in following the process of testing each spot and recording it accurately on the data collection sheet.

### Learner Profile: Jacob

"Jacob" had a medical diagnosis of attention deficit hyperactivity disorder. He had difficulty maintaining focus and cooperating with others. He tended to rush through his work and not pay close attention to details. He sped through the work to be able to read a favorite book. Science was one of Jacob's favorite subjects, so he tended to be more engaged and focused. He was a reluctant writer who struggled in organizing his ideas and recording them and whose handwriting was difficult to read.

Throughout the learning cycle, I implemented a variety of solutions to support Jacob in cooperating with others, attending to details, and accurately recording his results and ideas:

- In the Engage phase, I had Jacob verbally tell me his answer to the friendly talk probe to rehearse it before he wrote it in his science student packet. This allowed him to construct his explanation orally and add details before writing. Before rehearsing his answer with me, he had only written "#2." I asked him to explain why he chose that response and to add that in his student packet. Jacob then added that he agreed with student #2 "because it is not metle." He also added, "This was sooo Fun!!!"

- During the Explore phase, Jacob was eager to speed through the testing process. Mary was in his group, so I regularly visited the group to review the teamwork expectations of sharing materials, using kind words, and solving problems quickly. Teams rotated the recording sheet. By the third evaluation round at the end of the lesson, Jacob was the recorder. He was supposed to record what the team agreed on in regards to their choices during the science exploration. Instead he quickly drew smiley faces all down the sheet for round 3. His team disagreed with him, and he became upset as they reminded him that he had not shared the materials or used kind words and that the team had not solved the problems quickly. He hastily scribbled out the happy faces and turned them into sad faces. When I noticed the conflict over the reflection, I joined the group to listen to all team members and support Jacob as he changed the sheet. This minimized threats for Jacob.

- The Evaluate phase was structured in partnerships for the entire class to increase cooperation and to give students more hands-on opportunities as compared to groups of four. To support Jacob, I specifically chose a partner who would be patient if he chose to take a leading role and who would help Jacob attend to the details of testing the mystery boxes and recording his ideas to determine the material.

## Questions to Consider

➢ To what extent did the activities in this lesson align with the purpose and intent of each phase of the 5E Learning Cycle? Could you envision other activities that would be appropriate for each phase?

➢ Were you able to follow the sequence of activities and the ideas that students developed in the lesson? How did the storyline of the lesson progress?

➢ In what ways was the teacher able to assess students during each phase of the lesson? How did she respond to the difficulties she identified in students' understanding? How might you have responded?

➢ In what ways did the solutions that the teacher identified meet the needs of the specific students spotlighted in this vignette? In what ways did they benefit all students? Could you think of other solutions you might use for your own learners?

## References

Beaty, W. J. n.d. Children's misconceptions about science. Retrieved June 8, 2018, from *http://amasci.com/miscon/opphys.html*.

Hager, T. 2006. May the force be with you. In *Seamless assessment in science*, ed. S. K. Abell and M. J. Volmann, 61–70. Portsmouth, NH: Heinemann.

Keeley, P. 2008. *Science formative assessment: 75 practical strategies for linking assessment, instruction, and learning*. Thousand Oaks, CA: Corwin Press.

NGSS Lead States. 2013. *Next Generation Science Standards: For states, by states*. Washington, DC: National Academies Press. *www.nextgenscience.org/next-generation-science-standards*.

# CHAPTER 7

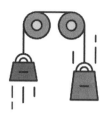

# You Really Dropped the Ball

*Mahaley Sullivan*

In this chapter, the learning cycle represents the first lesson in a fifth-grade unit about forces, which engages students in developing an understanding of how gravity affects objects. Students build this understanding through investigating falling objects that do and do not hit the ground at the same time when dropped from the same height, while figuring out how an astronaut on the Moon can similarly drop a hammer and a feather and have them hit the Moon's surface at the same time. The conceptual storyline (see Figure 7.1, p. 144) that helps students understand these phenomena builds over several days.

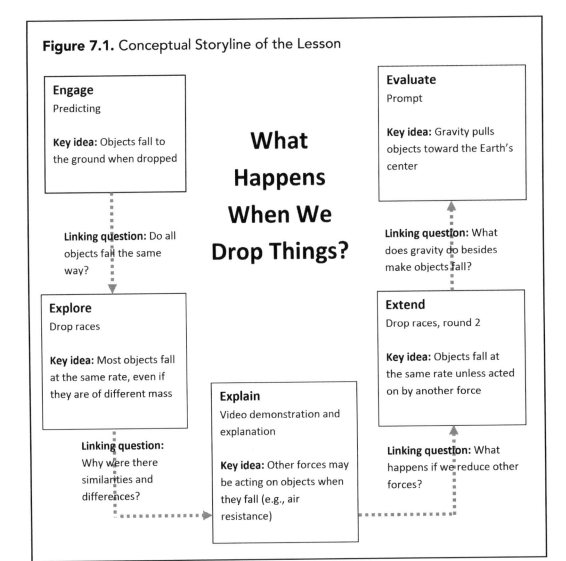

**Figure 7.1.** Conceptual Storyline of the Lesson

# What Happens When We Drop Things?

**Engage**
Predicting

**Key idea:** Objects fall to the ground when dropped

**Linking question:** Do all objects fall the same way?

**Explore**
Drop races

**Key idea:** Most objects fall at the same rate, even if they are of different mass

**Linking question:** Why were there similarities and differences?

**Explain**
Video demonstration and explanation

**Key idea:** Other forces may be acting on objects when they fall (e.g., air resistance)

**Extend**
Drop races, round 2

**Key idea:** Objects fall at the same rate unless acted on by another force

**Linking question:** What happens if we reduce other forces?

**Evaluate**
Prompt

**Key idea:** Gravity pulls objects toward the Earth's center

**Linking question:** What does gravity do besides make objects fall?

## LESSON VIGNETTE

I began planning this learning cycle by thinking about the standards that I needed to teach and the students within my room, and I mapped out a conceptual story-line. For this lesson, I focused on teaching about how the gravitational force exerted by Earth on objects is directed toward Earth's center.

As I mapped out activities and chose phenomena that aligned with this idea, I kept in mind that I would have to adjust these things depending on the needs of my learners. Before attending the Quality Elementary Science Teaching (QuEST) program, I would have started with the activities I wanted to include, but admittedly this didn't always result in a coherent conceptual storyline.

I find that using the conceptual storyline graphic organizer helps me map out my thinking and ensures that I am sticking to the same concept and developing the idea in-depth through the activities and phenomena I include in the lesson. I often share this document with my colleagues at school and have them double-check that it flows. If it doesn't make sense to the teacher, how can it make sense to students?

See Table 7.1 (p. 146) for alignment of this lesson to the *Next Generation Science Standards* (*NGSS*; NGSS Lead States 2013).

| TABLE 7.1. *NGSS* Alignment | |
|---|---|
| **Connecting to the *NGSS*—Standard 5-PS2-1: Motion and Stability: Forces and Interactions** | |
| *www.nextgenscience.org/pe/5-ps2-1-motion-and-stability-forces-and-interactions* | |
| • The chart below makes one set of connections between the instruction outlined in this chapter and the *NGSS*. <br>• The materials, lessons, and activities outlined are just one step toward reaching the performance expectation listed below. | |
| **Performance Expectation 5-PS2-1.** Support an argument that the gravitational force exerted by Earth on objects is directed down. | |
| **Dimensions** | **Classroom Connections** |
| *Science and Engineering Practices* | |
| *Engaging in Argument From Evidence* <br><br> Support an argument with evidence, data, or a model. | Students use their data to make a claim about what causes objects to fall at a similar or different rate. They develop a model of "air resistance" as a force that acts in the opposite direction of gravity. |
| *Disciplinary Core Ideas* | |
| *PS2.B: Types of Interactions* <br><br> The gravitational force of Earth acting on an object near Earth's surface pulls that object toward the planet's center. | Students explore dropping different objects and observe how they fall toward the ground. |
| *Crosscutting Concepts* | |
| *Cause and Effect* <br><br> Cause and effect relationships are routinely identified and used to explain change. | Students explore what happens to the time it takes a piece of paper to fall when they change its shape. |

## Engage Phase

I began my learning cycle with a simple question—what do students think would happen if I were to drop my keys? Students couldn't believe that I didn't know the answer to this, but I asked them to humor me and talk to their shoulder partner. I listened to the conversations and, where necessary, prompted students to share their ideas, then went to the front of the classroom and called back their attention. I said, "I heard the majority come up with an answer right away. Let's double-check that you're right, because with science we always want to test our ideas."

I dropped my keys several times in different ways, each time asking students to point out where they thought the keys would fall. "You've done a great job *predicting*," I began. "So now I'd like you to try *explaining* why the keys fall." I provided the students with sticky notes on which they could either draw or write down an explanation. When they were finished, I asked them to place these notes on the chart paper on our easel, so we could reference them later.

**UDL Connection**

*Principle II: Action and Expression*
*Guideline 5: Expression and Communication*
*Checkpoint 5.1: Use multiple media for communication*

After the students started working on the task, I circulated around the room to check in with specific students and offer additional support that they may need to complete the task. I first made my way to Donald, a student who had processing and writing difficulties. I repeated the question to Donald, "Why did the keys fall to the ground?" He paused and went "Hmmm" while looking up, which signaled to me that he was still processing his ideas, so I told him I would be right back to check in on him.

Next, I visited my student James and asked him what he was thinking. James had a learning disability in language and a limited vocabulary, so I offered to be his scribe. James responded, "I already have a picture, but I don't know what to label." His picture showed the keys with an arrow on top and below pointing in the same direction (downward). I then asked him what he thought happens, and I was told that something pushes on it, which is why he used the arrow on top. He also told me he had the other arrow because that was the direction it falls. I noticed he was using arrows to represent the force as in a physics diagram for the pull of gravity, but he was showing the direction of movement of the keys as opposed to the push of air in the opposite direction to gravity. I asked him what he thought the thing pushing on the keys might be, and he told me he thought it was air. I instructed him to then label the arrow he had on top of the keys as air.

I then turned back to Donald and asked if he was ready. He did a big head nod, and I asked him to tell me what to write. He said, "Umm, I think it's … it's ummm … it's gravity pulling on it to make it fall." I wrote on his sticky note, "Gravity pulling on the keys makes them fall." I then had him read it over and asked him if I needed to change anything. He said no, so I told him to place it on the chart paper.

> **Teaching Tip:** As I circulate around the room, I have specific students in mind that I want to check in on and offer support.

By the time I had finished with Donald, the rest of the class was done. So I pulled them back together, and we read over the sticky notes together. The majority of the class believed that it was gravity pushing or pulling on the keys to make them fall. A few said that air was pressing them down. While air does push on falling objects, that push is actually in the opposite direction to the pull of gravity. "What if I had a keychain with twice as many keys?" I asked.

Students expressed with confidence that they would still fall when dropped, though some thought they would fall faster and others thought it wouldn't make a difference. I did a mental check of my plans to consider how these ideas would be elaborated on or, in the case of misconceptions about the direction that air pushes on the keys, confronted with evidence in the activities. Because this was simply a time to examine students' current ideas, I told them we would look back at these ideas later in the week after we've had a chance to investigate.

## Explore Phase

For this portion of the lesson, I decided to do "drop races" in which students would observe whether objects dropped at the same time from the same height hit the ground at the same time. I made a bag of materials with a tennis ball, large marble, 2.5-inch foam ball, bouncy ball, ping-pong ball, toy helicopter spinner, golf ball, feather, and flat sheet of white copy paper for the students to do drop races with. I picked this variety of objects because I knew a few would be subject to greater air resistance, and fall more slowly, while others would show that objects fall at the same rate even if they have different mass. I picked the large marble because it was about the same size as the bouncy ball, and I picked the foam ball because it was about as big as the tennis ball. Thus, the air resistance would be similar for both, even though their mass differed. (Please see the "Materials and Safety Notes" box.)

---

### Materials and Safety Notes

**Materials**

| | | |
|---|---|---|
| Tennis ball | Ping-Pong ball | White copy paper |
| Large marble | Toy helicopter spinner | Digital scale |
| Foam ball | Golf ball | Fabric measuring tape |
| Bouncy ball | Feather | |

**Safety Notes**

1. All involved must wear indirectly vented chemical splash goggles or safety glasses with side shields during all phases of these inquiry activities (setup, hands-on investigation, and take-down).

2. Direct supervision is required during all aspects of this activity to ensure that safety behaviors are followed and enforced.

3. Make sure that any items dropped on the floor or ground are picked up immediately after working with them—a slip/trip fall hazard.

4. Ensure that all fragile items have been removed from the activity area to prevent breakage and potential injury.

5. Follow the teacher's instructions for returning materials after completing the activity.

6. Wash your hands with soap and water after completing the activity.

---

I gathered the students on the carpet and explained that they would be examining whether objects fall at the same or a different rate. I showed them a set of the objects they were to test and asked them how they were the same or different. Students recognized that they were different sizes and, from prior experience, that some had different mass. "Would that make a difference in how they fall?" I asked. Students disagreed on that, and so we agreed that recording information about the size (diameter for round objects) and mass would be important for interpreting our results. I assigned this task and offered them a choice in making their own data table or using one I had premade. I placed all materials and the copies of data tables on one table, so that the groups could gather what they needed and get started.

**UDL Connection**
*Principle III: Engagement*
*Guideline 7: Recruiting Interest*
*Checkpoint 7.1: Optimize individual choice and autonomy*

I walked around to observe students as they examined the objects. We had already focused on measurement as part of our earlier unit on matter, so I was pleased to see groups immediately set to work on placing different objects on the digital scales. I noticed some using another object to keep the balls from rolling off the scale and then subtracting the mass of that object to find the mass of the ball. It was exciting to see them using a strategy that they had developed four months before! For diameter, students similarly were using fabric measuring tapes as they had previously learned.

The students were strategically placed in groups to avoid potential problems, and I put Donald and James in the same group so that I could assist them at the same time, checking in to be sure they were recording the information in their notebooks. In addition, Donald would often lose focus while doing a task and not finish it. This would cause him to miss key information, so I put him with a student who was very adamant about having everything completed and written down accurately. She would keep track of what Donald was doing and say in a nice tone, "Donald, I think you missed a number." He would then refocus and ask what to write in the table.

**UDL Connection**
*Principle III: Engagement*
*Guideline 8: Sustaining Effort and Persistence*
*Checkpoint 8.3: Foster collaboration and community*

I also noticed while walking around that one of my groups, on their own, thought to split up the length of the stick and blade of the helicopter spinner and that they should tell both how long and how wide the paper was. I asked them why they had done that, and they said because there were two parts that needed to be measured. I then paused to get everyone's attention and had the group share what they were doing. I did this because it made a group shine and because I wanted to be sure all groups were taking this into account in their measurements.

After the students had completed their observations of the objects, I gave each student a handout with a data table (see Figure 7.2) showing possible pairings of the objects that could be compared in terms of how they fall. I explained that the students would have two objects "race" to the ground when dropped, and I asked them to review the list and indicate individually whether they thought one object would hit the ground first, or whether they would hit the ground at the same time. I went over an example by projecting the table onto the board with my document camera. I then gave this task to the students to complete as the final task for that day's session.

The next day, students were excited to test their predictions and to hold the drop races. Before they began, however, we planned together how to ensure that the races were fair. I began by asking a volunteer to drop one object while I dropped another. As the volunteer held out the object, I made sure to hold mine higher—much to the objection of the students! I asked them to generate a list of "rules" for the drop races. The students agreed that the objects should be dropped at the same time, in the same way, from the same height. I pointed out several "drop zones" that I had created around the room for students to use, and I asked the students what kinds of safety precautions they might need to take (e.g., staying out of the drop zone when objects are being dropped, wearing safety glasses).

**Figure 7.2.** Sample Data Table Used to Record Predictions for the Drop Races

| Object 1 | Object 2 | Object 1 will hit the ground first | Object 2 will hit the ground first | Both will hit the ground at the same time |
|----------|----------|-----------|-----------|-----------|
| Ping pong ball | Golf ball | | | |
| Ping pong ball | Feather | | | |
| Ping pong ball | Foam ball | | | |

I then demonstrated with four volunteers what it would look like to drop two objects, so that the students would know how to drop, where the observers would need to be (on their bellies a few feet back from the drop zone), and when the timer should start. Early on in the school year, I learned that if I allowed this class to pick who did the various parts, the students would argue and fight over the different tasks instead of working together. To help alleviate this issue, I assigned each group member to have a certain task, and to prevent dissatisfaction with the job task, I rotated tasks for each drop race.

I talked through an example of how it would work before starting. For example, first I might be the person who drops the objects. Then, after I did a round of dropping objects, I would become the person who times the objects. Then, for the next round, I would be one of the eye-level observers. For the last round, I would be the other eye-level observer. I gave all of them a premade chart of the races so that, particularly for students who had difficulty making their own organizational tools, they would not miss one and would have a way to record the data.

**UDL Connection**
*Principle III: Engagement*
*Guideline 8: Sustaining Effort and Persistence*
*Checkpoint 8.3: Foster collaboration and community*

**UDL Connection**
*Principle II: Action and Expression*
*Guideline 6: Executive Functions*
*Checkpoint 6.3: Facilitate managing information and resources*

I then told the students that I would be walking around, taking videos of some of the drops, and assisting. I began with Donald and James' group first so that if they needed help, I could support them. Donald had trouble with handwriting so that it took him longer to write, so I told him that I just wanted him to write down whether one object dropped first or at the same time as the other object. I told him to ignore the column for the time. He

> **Teaching Tip:** The first time that I did this activity, I wanted students to time the drops. However, many of the objects fell so quickly that students were unable to get accurate times and became frustrated. They were, however, able to notice more accurately which object hit the ground first—so I modified the activity to focus on that.

did a gigantic head nod with a wide grin across his face. I then got down on my stomach and took a video with my iPad of their first drop race.

**UDL Connection**
*Principle III: Engagement*
*Guideline 8: Sustaining Effort and Persistence*
*Checkpoint 8.2: Vary demands and resources to optimize challenge*

> **Teaching Tip:** While students have lots of experience working in groups, even the best-laid plans can go awry, and I've learned over the years that I need to do a lot of modeling and providing prompts at the beginning of our year about how to work effectively with others during science and how to handle disagreements. I have the students create posters describing what to do when they disagree, so that they can refer back to the posters when problems arise.

Noticing that some students were off-task, not following instructions, and arguing within their groups, I stopped the class to revisit our expectations. I picked two student volunteers (one who I observed had been arguing and one who had not). I then gave them a scenario in which one of them thinks that the marble fell first and the other thinks that the golf ball did. Before responding, I reminded the students to use the "I" statements we had previously learned as a way to resolve disagreements. Barry began, and he said, "I think that the marble fell first because I heard a ping. Why do you think that the golf ball fell first?" Then the other student said, "I see why you picked the marble, but I saw the golf ball hit the ground first." I then asked them if they had an idea of how they could further test which fell first. Barry said, "We could videotape it ...," and then paused and asked, "Is that why you were taping?" I nodded.

I then asked if there was anything else that they noticed we needed to work on in the groups. Someone piped up and said, "People were not writing down information." I told them I noticed that too and asked them why it's important to write information down. They told me that it helps us draw conclusions later. Noticing we were near the end of our time for science, I added, "It also helps us remember tomorrow what we did today!" To wrap up, I asked each student to jot down one thing he or she would like to do tomorrow to help the group work well together.

The next day, I didn't go immediately intervene with a group, but I walked around and hung back to observe. I noticed students having different interpretations of which object hit the ground first, but instead of arguing, they decided to try dropping again, or they asked me to videotape their drop. I circulated among the groups, checking in to see if there were any objects they had difficulty comparing, and I supported them in retesting or taking video. Once they had collected all of their data from the drop races, I asked them to make a generalization about how objects fall.

For Donald and James, I gave their group a prompt to assist them: "Objects typically hit the ground _____ because _____." James then said, "But not all did hit at the same time!" I asked him what the things had in common that didn't fall at the same time as the other objects. He had a light-bulb moment at that time and said that made a lot more sense.

> **UDL Connection**
> *Principle II: Action and Expression*
> *Guideline 5: Expression and Communication*
> *Checkpoint 5.2: Use multiple tools for construction and composition*

After everyone was finished, I pulled them together and had them sit in a large circle next to their group members to share what they had found. All the students noticed that the majority of the drop races were a tie with the exception of the toy helicopter spinner, paper, and feather. Some groups also included the foam ball among those objects that hit the ground at a different time than other objects, and we agreed we should check our videos to help resolve this discrepancy. As we watched a few of the slow-motion videos of races between other objects and the foam ball, students noticed that the majority of the time it hit at the same time as the other objects. Only in one instance (against the golf ball) did it ever so slightly not hit at the same time. The difference was small enough that students thought it could have been dropped at a slightly different time.

I then asked the students to come up with a general rule to describe how objects fall, and I told them to talk to one of the people sitting next to them. As I walked around, I heard the students saying almost the same exact sentence: "Most objects on Earth fall at the same rate or very close to the same rate." If this is true, then what caused some objects to take longer to fall? I left students to wrestle with that question until the next session.

## Explain Phase

Although students had already begun to generalize from patterns in their data, an important part of the Explain phase involves checking ideas with other sources and solidifying understanding about key concepts. For this phase I identified two videos for this purpose. "Crash Course Kids" provides a student-friendly definition of gravity and how it affects objects, particularly in the first portion through the time of 6:25 (*www.youtube.com/watch?v=EwY6p-r_hyU*). I believed this would provide students with a visual representation and put the idea into kid-friendly terms that would be accessible to everyone in my classroom, particularly those learners who had difficulty reading text. It would also help them consider how their own findings are consistent with the examples in the video.

**Teaching Tip:** In this video segment, the narrator uses a rubber band and a tennis ball to model gravity. Having these materials on hand for students to try for themselves can support their interpretation of this model.

I also chose a video of astronaut David Scott dropping a hammer and a feather on the Moon (*www.youtube.com/watch?v=5C5_dOEyAfk*). I picked this video because, unlike in my students' tests, it shows that even a feather and a hammer will hit the ground at the same time—something that was inconsistent with their results, and as such it would motivate them to understand the why of key concepts.

I began my lesson by gathering the students on the carpet and asking them to turn to a neighbor and remind each other what they had figured out yesterday and what they were still wondering about. The majority of students were realizing that most objects fall at the same rate and they wondered why not all did, or they were surprised by the results because they had expected that heavy things would fall faster. I commented on this as I gathered the students back together, and I told them that they could check what they had found against what scientists have discovered, to help make further sense of their ideas. After we watched the video, I asked the students to turn and talk to their neighbor about what gravity is and how that might help explain what we found. Students felt more confident in their findings and their ideas about how gravity was responsible for this, but they were still puzzled by why not all objects tied in the drop races.

**Teaching Tip:** I noticed a few of the kids were starting to get restless at this point, so instead of going straight to the next video, I decided to take a break. I had them get up and move around the room and do the Electric Slide with "Go Noodle" (*www.gonoodle.com*).

We then watched the video of the astronaut dropping a hammer and a feather to help students think about objects that *didn't* fall at the same time. The students were expecting the hammer to fall more quickly, and I heard a few of them say that it didn't fall at the same time. We ended up rewatching the video until we had a consensus that they dropped at the same rate. I then told the students to talk to their shoulder partner: "Why do you think the feather falls at the same rate as the hammer on the Moon, but not on Earth?" I knew that they had previously learned about this in third grade during their weather unit, and they had touched on air resistance during their motion unit in fourth grade, so I was hoping they'd draw on this prior knowledge.

I circulated around to hear students' conversations, making my way to James and his partner first. James already had a wrinkle in his eyebrow and was really thinking. I asked him what he remembered from yesterday, and he said that the feather was what fell slowly. I then asked, "What does the Earth have that the Moon does not?" His partner then said, "Ooooh, I know … air, because astronauts have to have a helmet to breathe." I encouraged them to talk further about why that might make a difference, then moved on to another group.

I made my way to Donald, after noticing his partner was doing all the talking. "Donald, what do you think about why the feather fell at the same rate as the hammer on the Moon but not on Earth?" He looked at me and said, "Air resistance." I asked him to tell me more, and he offered, "The air keeps it up."

We came together as a whole group and reached a consensus that air made a difference in how some objects fell, specifically air resistance. I asked the students to go back individually and rewrite their statements from the previous day about how objects fall, to incorporate ideas about gravity and air resistance into their explanations. As most students began writing, I checked in with Donald and James, for whom I knew writing poses a challenge, often resulting in incomplete ideas or even refusal to write.

> **Teaching Tip:** In hindsight, I would not suggest using the toy helicopter spinner, because while it did show that not all things drop at the same rate and was closer in weight to some of the other objects, there was no way students could change it without adding or taking something away. Instead I would use another object that the students could easily change, such as a plastic bag.

To get Donald started, I asked him to tell me what he would change from yesterday. He said, "Because of gravity all objects fall to the ground, but air resistance slows some." I replied, "Wow … you included both ideas!" and his expression became a big, wide smile. "Go ahead and write this down," I said. James, however, sat with his brow creased, a signal to me that he was struggling, so I opted to interview him to better understand how his ideas had changed. I asked what he had written last time. He said, "Objects typically hit the ground at the same time because most races did. A few objects didn't and they were lighter." I encouraged him to think, "Why did the objects typically hit the ground and not float?" When he replied "gravity," I asked, "How does gravity do that?" He responded that gravity pulls objects to the ground. Then I asked him why several of the objects didn't fall at the same time. He said they were slowed by the air pushing against them.

**UDL Connection**
*Principle II: Action and Expression*
*Guideline 5: Expression and Communication*
*Checkpoint 5.1: Use multiple media for communication*

## Extend Phase

I continued the lesson in the next session by asking students to recall why certain objects such as paper took longer to fall, and I asked whether they thought it would be possible to do something to one of the objects (paper, helicopter, or feather) to change how it falls. They eagerly said "Yes," anticipating what they would be doing next! I explained that they could *change* the object, but they could not *add* anything to the object. I then charged each group with deciding what they wanted to change about the object, and how they would test the effect of this change. I asked them to write down their plan in their notebook or draw it out with labels, and then I would give them the materials to test it out. I also told them that because they would all be testing different things, each group would need to make a poster to share their findings with the other groups.

**Teaching Tip:** In previous years, I had extended the lesson further using an online simulation that showed how gravity on the Moon and Earth differ. This version of the lesson progressed more naturally from the previous phases based on students' developing ideas and questions.

I began walking around as the students began brainstorming. I heard them all trying to think how they could change the feather, the toy helicopter spinner, and the paper. All of the groups immediately decided that the spinner wouldn't be easy to change because of its rigid shape—it would break. I had expected some groups to choose the feather, but all settled on paper.

I overheard some arguing happening in Barry's group, and I checked in on them. Barry wanted to crumple the paper into a ball as tight as possible. Amy wanted to fold the paper until it couldn't be folded anymore. I asked both students to explain their choice. Barry said he wanted to crumple it into the ball because it would make the paper into a similar shape as the other objects we had dropped, whereas folding wouldn't. Amy said that she thought folding the paper would allow more people to participate—letting each person make a fold. I asked them if they would like to try both. "Yes!" they both agreed. Barry then asked me if he could add something else. I asked him to elaborate. He told me he wanted to compare the length of time it took for the paper to fall by itself flat and then again after he made the ball, because it may not still hit the ground at the same time or close to it. I paused everyone so he could share his idea and have his group respond. The others agreed it was a good idea because it would let them know if they had actually changed the amount of air resistance.

**UDL Connection**

*Principle III: Engagement*
*Guideline 8: Sustaining Effort and Persistence*
*Checkpoint 8.2: Vary demands and resources to optimize challenge*

After they had finished, I provided the students with materials to make their posters. The students all drew pictures of what they did with the exception of one group who used writing only. By not restricting how they communicated their ideas, I was able to meet multiple students' needs such as students who were reluctant writers but who would draw to communicate their ideas, and I kept them engaged. After all of the posters were ready, we hung them around the room for a gallery walk.

**UDL Connection**

*Principle III: Engagement*
*Guideline 7: Recruiting Interest*
*Checkpoint 7.1: Optimize individual choice and autonomy*

Before we began the gallery walk, I supplied each group with sticky notes and reviewed expectations for feedback. I explained that one constructive feedback strategy is to ask a question about anything that is not clear. We agreed that this would be more helpful than stating, "This doesn't make sense." I also told them that they should give positive feedback to help groups understand what they did well; however, it couldn't simply be writing "Good job."

To help students generate informative feedback for their peers (and as a way to show them what questions they could ask themselves when reflecting on their work), I provided the students with the following prompts:

- "I like how you _____, but did you think about _____?"
- "I really enjoyed _____ on your poster because _____."
- "Why did you pick this way of doing _____?"
- "I really liked how you thought about _____."
- "Did you think about trying _____?"

I reminded the students to point out at least one thing they think each group did well and one thing they could improve. I then had students go around and put comments on the posters.

> **UDL Connection**
> *Principle II: Action and Expression*
> *Guideline 6: Executive Functions*
> *Checkpoint 6.4: Enhance capacity for monitoring progress*

After the groups had finished providing their feedback, I asked the class to brainstorm things they felt made effective posters. I recorded these on a chart with the intent of keeping this for future use when making posters. Among the characteristics that students identified were organized sections, labeled drawings, and supporting details. I then had the groups go back and add to or change their poster based on the feedback.

## Evaluate Phase

I had less flexibility in designing this phase of my lesson, as my district requires that teachers use the same summative evaluations across grade levels and buildings for science. For this unit, that consisted of a writing prompt that asks the students to give two examples of how gravity affects them and to explain how the example demonstrates that. It was not a "three-dimensional assessment." I had learned in QuEST that it is essential to offer a choice and to adapt assessments so that they are accessible to everyone. As a result, though I couldn't change the assessment task, I gave my students the option of making a video to communicate their answers, drawing models, verbally responding, or writing their ideas down. Many students chose to write; however, for those who chose to respond orally, I transcribed their explanations and noted that it was provided to me verbally.

> **Teaching Tip:** Because this was the students' first experience making posters, we used this activity as a way to build their understanding of how to communicate effectively in this format by generating their own criteria. The chart we created could be revisited and added to as we made posters in different contexts.

> **UDL Connection**
> *Principle II: Action and Expression*
> *Guideline 5: Expression and Communication*
> *Checkpoint 5.1: Use multiple media for communication*

Several of my students drew a diagram similar to the penguin example from the "Crash Course Science" video, substituting themselves for the penguin and using arrows to show that gravity was pulling them toward Earth's center. I also had several kids draw a picture of them dropping something and saying that gravity caused it to fall down, because gravity pulled it toward Earth's center. I had a few who showed something falling on their heads (ouch!) and explained that gravity pulled it down on top of them. Written responses included similar scenarios to these three.

Ideally, an assessment should bring the conceptual storyline full circle by returning to the ideas you elicited from students in the Engage phase. This prompt was appropriate in that regard. As I reflected on students' responses, however, I realized that this particular assessment did not address the idea of air resistance as a force that opposed gravity. Therefore, I decided to revisit this idea in the next learning cycle of the unit.

## Unpacking UDL: Barriers and Solutions

As teachers, we are often asked to reflect on our lessons and to think about what we would do differently and change if we were to do it again. The thing I liked about the approach to framing our instruction that was introduced to us in the QuEST program is that it required us to anticipate potential problems *before* we faced them, so that I felt less overwhelmed when things did go wrong. I already had an idea of what to fix or how to fix it because of the Universal Design for Learning (UDL) guidelines. I cannot stress enough how often I would look at those when planning, as well as when unanticipated barriers emerged.

Table 7.2 (p. 160) summarizes the UDL principles, guidelines, and checkpoints that I applied when designing the activities for each phase of the 5E Learning Cycle lesson to meet the *general* needs of the learners in my classroom. Following that are examples that illustrate how I identified barriers and strategized solutions to meet the *specific* needs of some of the learners in my classroom. You may recognize these students from my vignette!

| TABLE 7.2. UDL Connections | |
|---|---|
| **Connecting to the Principles of Universal Design for Learning** | |
| *Principle I. Representation* | |
| *Guideline 2: Language and Symbols* | |
| Checkpoint 2.5. Illustrate through multiple media | Some students struggle to access ideas through text. During the Explain phase, videos were used to illustrate key concepts. |
| *Guideline 3: Comprehension* | |
| Checkpoint 3.3. Guide information processing and visualization | During the Explain phase, videos were used to provide interactive models for learning for those who needed additional time to process information for understanding and recall. Questions following each video were asked to guide students' understanding and recall. |
| *Principle II. Action and Expression* | |
| *Guideline 5: Expression and Communication* | |
| Checkpoint 5.1. Use multiple media for communication | To help students who struggled to write, options were provided during the Engage and Evaluate phases to communicate ideas via writing, drawing, or other media. |
| Checkpoint 5.2. Use multiple tools for construction and composition | Some learners struggled to start writing or organize their ideas, so during the Explore phase, a sentence starter was provided. |
| *Guideline 6: Executive Functions* | |
| Checkpoint 6.3. Facilitate managing information and resources | All students were provided with a premade chart to record data to ensure that the students, including those who had difficulty writing, could manage information. |
| *Principle III. Engagement* | |
| *Guideline 7: Recruiting Interest* | |
| Checkpoint 7.1. Optimize individual choice and autonomy | For some students, providing options on how to record data (on their own or with a premade table) can help facilitate data management while giving autonomy for task completion. |
| | During the Extend phase of the lesson, to maintain engagement and interest for communicating ideas learned, the students were provided choices of how to present their ideas. |

*(continued)*

| TABLE 7.2. UDL Connections (*continued*) | |
|---|---|
| **Connecting to the Principles of Universal Design for Learning** | |
| *Guideline 8. Sustaining Effort and Persistence* | |
| Checkpoint 8.2. Vary demands and resources to optimize challenge | For one learner, expectations for the amount of writing to do during the Explore phase were reduced to ensure task participation and completion. |
| | Students who are gifted often need more challenge to keep them engaged in learning. In the Extend phase of the lesson, an option to increase the challenge of the task was provided. |
| Checkpoint 8.3. Foster collaboration and community | Strategically organizing groups, such as during the Explore phase, can create opportunities for peer support and interaction as a way to keep all members on task. |
| | During the Explore phase, students were placed in groups with assigned roles and responsibilities to ensure equal participation and task completion. |

### Learner Profile: Donald

"Donald" was a student with multiple complex learning needs, and while he had an individual education plan for "other health impairment" and a learning disability in the area of writing (e.g., production of text and composition), he also had been identified as having behavior problems, major depression, and autism spectrum disorder. He went to the resource room with a special education teacher for writing, but was in the general education classroom for all other subjects. Due to significant family issues during the year, such as the loss of a family member, Donald's coping mechanism was to internalize it, and consequently he fell back on old habits of not communicating. This became a challenge in all academic areas including science.

To overcome these challenges, I often used the following strategies during any phase of the learning cycle:

- The option to respond and communicate ideas in a variety of ways
- Provision of data collection tables
- Reduced writing expectations
- A student partner to assist in task persistence and completion
- A scribe for writing tasks (student or teacher)

### Learner Profile: James

"James" was a student who had a reading disability. He performed academically on a second-grade level in reading and writing. For any activity throughout the lesson that required writing or organizing data, he would struggle.

To help decrease the impact of these challenges, I used the following strategies for James:

- Providing prompts such as sentence starters to develop explanations
- Providing data collection tables
- Checking in with the student to ask probing or clarifying questions to elicit student ideas
- Allowing the student to respond by drawing pictures or verbally, instead of writing it down (provide a scribe if it's required to be on paper)

### Learner Profile: Barry

"Barry" was a student who had been identified as being gifted and talented. He attended an extension program to help challenge him once a week with other gifted students from around the district. During science, and in other content areas, he had difficulty keeping his knowledge to himself (he wanted to shout out and give answers before others were ready for it). He also would take over for the group and do all the work. He had trouble taking others' suggestions and would argue over how to do something.

For any activity that required group work, Barry struggled to let the other classmates do the work or to share ideas appropriately. To overcome these barriers to learning, I did the following:

- Assigned tasks for each student to complete and rotated who did which job
- Allowed groups to try more than one way to solve a problem
- Allowed students to respond in a multitude of ways
- Pretaught strategies to resolve disagreements that might occur when working in small groups

## Questions to Consider

> To what extent did the activities in this lesson align with the purpose and intent of each phase of the 5E Learning Cycle? Could you envision other activities that would be appropriate for each phase?

> Were you able to follow the sequence of activities and the ideas that students developed across all five phases? How did the storyline of the lesson progress? In what ways did the teacher identify incoherence in the storyline? How might you adjust the activity to improve the conceptual coherence across all five phases?

> In what variety of ways was the teacher able to assess students' ideas during each phase of the lesson? How did this inform her instruction in this lesson? How did this inform her plans for the rest of the unit?

> In what ways did the solutions that the teacher identified meet the needs of the specific students spotlighted in the vignette? In what ways did they benefit all students? Could you think of other solutions that you might use for your own learners?

## Reference

NGSS Lead States. 2013. *Next Generation Science Standards: For states, by states.* Washington, DC: National Academies Press. *www.nextgenscience.org/next-generation-science-standards.*

# CHAPTER 8

# Save the Penguins!

**Betsy O'Day**

"How can we keep a penguin nest cool"? This chapter addresses that question with a lesson designed to incorporate engineering. It was taught at the end of a unit about heat energy for fourth graders. The students drew on knowledge they had gained throughout the unit to develop a solution to the problem. The conceptual storyline of the lesson (see Figure 8.1) is outlined on page 166.

**Figure 8.1.** Conceptual Storyline of the Lesson

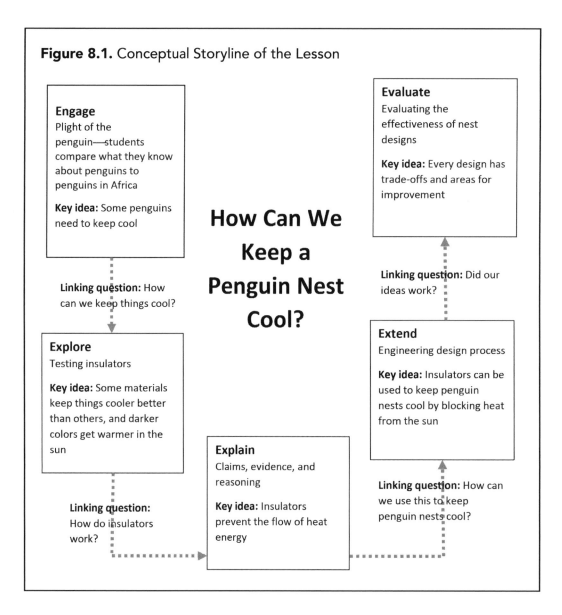

## LESSON VIGNETTE

As a teacher, I like to engage students in science that helps connect them to the real world beyond our classroom. After reading about an activity for third-grade students in *Science and Children* by Sheerer and Schnittka (2012), I was inspired to design this lesson, and I aligned it to the *Next Generation Science Standards* (*NGSS*). I had read about the plight of penguins in South Africa,[1] and I knew this would be a compelling phenomenon for my students. Many of them had grown up in rural Missouri and were fascinated by animals that they could only see in zoos, such as penguins. However, I anticipated that many of them would also envision cold climates—snowdrifts in the Antarctic—when they thought of penguins. African beaches and high temperatures wouldn't be the first thing that comes to mind! As a result of shortages of food, exposure to the sun, and predators, the population of African penguins is plummeting. Conservationists have developed nest boxes to address the problem. Because I teach a unit on heat energy focusing on heat transfer, I thought this would provide an ideal context for students to design their own solutions to the problem to understand how the nest boxes work!

The problem in the engineering design task (Engage phase of this lesson) was presented at the beginning of the heat portion of the energy unit, to set a purpose for learning. Students were asked to design a nest cover for African penguins that would keep them cooler for longer periods of time. This would keep them on their nests longer and lessen the potential for predation of the eggs. To understand how to design a solution, students needed to develop an understanding of heat energy and how heat energy is transferred. To evaluate their solutions, the students would need to identify what data to collect and how to collect them, as well as to devise ways to represent and analyze their data to compare solutions. This became the bulk of our unit.

The 5E Learning Cycle provided me with a structure to incorporate the science and engineering into a coherent conceptual storyline. Given the highly collaborative nature of the project and the challenges inherent in the tasks, Universal Design for Learning (UDL) provided me with a way to anticipate possible barriers and incorporate solutions in my planning for the lesson.

For this lesson's alignment to the *NGSS* (NGSS Lead States 2013), see Table 8.1 (p. 168).

---

1. See *www.sanbi.org/animal-of-the-week/african-penguin*.

| TABLE 8.1. *NGSS* Alignment | |
|---|---|
| **Connecting to the *NGSS*—Standard 3-5-ETS1: Engineering Design** | |

*www.nextgenscience.org/dci-arrangement/3-5-ets1-engineering-design*

- The chart below makes one set of connections between the instruction outlined in this chapter and the *NGSS*.
- The materials, lessons, and activities outlined are just one step toward reaching the performance expectations listed below.

**Performance Expectations**

**4-PS3-2.** Make observations to provide evidence that energy can be transferred from place to place by sound, light, heat, and electric currents.

**3-5-ETS1-2.** Generate and compare multiple possible solutions to a problem based on how well each is likely to meet the criteria and constraints of the problem.

**3-5-ETS1-3.** Plan and carry out fair tests in which variables are controlled and failure points are considered to identify aspects of a model or prototype that can be improved.

| Dimensions | Classroom Connections |
|---|---|
| **Science and Engineering Practices** | |
| *Analyzing and Interpreting Data* Represent data in tables and/or various graphical displays (bar graphs, pictographs, and/or pie charts) to reveal patterns that indicate relationships. Use data to evaluate and refine design solutions. | Students analyze data in tables and graphs to identify materials best suited for use as insulators. Students analyze data from their tests to evaluate their nest box designs. |
| **Disciplinary Core Ideas** | |
| *ETS1.C: Optimizing the Design Solution* Different solutions need to be tested in order to determine which of them best solves the problem, given the criteria and the constraints. (3-5-ETS1-3) | Students compare the effectiveness of different designs in keeping penguin nests cool. |
| **Crosscutting Concepts** | |
| *Influence of Science, Engineering, and Technology on Society and the Natural World* Engineers improve existing technologies or develop new ones to increase their benefits, decrease known risks, and meet societal demands. | Students learn how engineers can use scientific ideas to help protect the African penguins from extinction. |

## Engage Phase

I started the unit by asking my class the following question: "What do you know about penguins?" Immediately, students seemed to perk up! I asked them to turn and talk to their shoulder partner. Some had seen penguins at the St. Louis Zoo, while others based what they knew on movies, television, or books. Because students typically think that penguins only live in polar environments, I showed them an excerpt of a BBC video of African penguins at Boulders Beach from YouTube (*https://youtu.be/mFd4Ibqi3Ig*) and said, "How are these penguins the same and different from what you and your partner discussed? Turn and talk again." I knew I had captured their attention as students expressed surprise at seeing penguins on a sunny beach!

> **Teaching Tip:** When students are eager to share what they think, calling on one student at a time can be counterproductive. Having them turn and talk engages everyone in expressing their ideas versus only a few.

I confirmed for the students that, indeed, some penguins live in temperate and tropical climates. This poses different problems for them. One of those problems is keeping cool. African penguins used to nest in guano, but humans removed all of the guano to use for fertilizers. Now they nest wherever they can find a place. Very often that is out in the open. When the penguins get too hot on land, they use the ocean to cool off. This leaves the nest open to predators like mongooses, snakes, sacred ibises, and kelp gulls. Once the eggs hatch, heat also affects penguin chicks when the adults are off the nest looking for food. Sometimes the chicks are left so long that they die from the heat and starvation. Often predators take them. There were 150,000 pairs of penguins in 1956 and only 26,000 pairs in 2009 (Braun 2010).

One of the methods being used to address this problem is the creation of nest covers that keep the penguins cool longer. I explained to my students that in this unit, they were going to be learning some science that would help them design and test nest covers to keep penguins cool.

## Explore Phase

Because our Engage phase activity had kicked off the unit, we picked back up on our storyline for the lesson at the end of the unit, after students had conducted a series of learning cycles to build an understanding of the following:

- Temperature is a measurement of heat energy in an object.
- Heat energy moves from warmer objects to cooler objects until they are both the same temperature.

- Conductors are materials that gain or lose heat easily. Insulators are materials that slow the transfer of heat.
- Materials can keep things warm by trapping heat.

Because some of my students often don't see connections between what they already know and the current task, I purposefully activated this prior knowledge and reconnected it to the plight of the penguins by asking students to think about why a penguin's nest might heat up during the day. I asked them to think about their own experiences as well. Some students drew on their experiences when visiting a beach (the sand is really hot!) or standing on concrete versus grass. Others brought up wearing black clothing versus light clothing on a sunny day— and how they felt warmer in the sun wearing black. Still other students made connections between the idea of a "box" and a "cooler"—the kind they use to keep things cool on a camping or boating trip.

**UDL Connection**
*Principle I: Representation*
*Guideline 3: Comprehension*
*Checkpoint 3.1: Activate or supply background knowledge*

Drawing on their examples, I asked students whether they thought the color of the nest box would matter. While the class leaned toward "Yes," some students were hesitant to take a position without testing further. I also asked the students to consider what coolers are made of—which resulted in a variety of answers (e.g., polystyrene foam, plastic, metal). Similarly, I asked students whether they thought the type of material we use for the nest box would matter. The consensus was yes, though the students expressed uncertainty about which would be the best choice. I asked them for some suggestions of materials that might work. Luckily, I was able to provide opportunities for students to gather data and test their ideas in our next session! See the "Materials and Safety Notes" box.

I'm fortunate enough to have temperature probes to use in my classroom, and when we resumed our exploration, we used these to compare the temperature changes of two paper heat pockets (one that is black and one that is white) when left under a heat lamp. The heat pockets were created from cardstock, and the temperature probe was placed inside each pocket (see the photo on the next page).

While the students were waiting for the data from that investigation to be generated over a 30-minute time period, I drew their attention to six bottles of water wrapped in different materials (cotton, wool, foil, bubble wrap, paper towel, and nothing). I explained that the bottles were cooled in the refrigerator overnight, then set out in the same location with temperature probes collecting data every five minutes for three hours (see Table 8.2, p. 172).

## Materials and Safety Notes

**Materials**

Temperature probes

Heat pockets made of cardstock (black and white)

Bottles of water

Cotton

Wool

Foil

Bubble wrap

Paper towels

Ice-cube penguin

Lamps

**Safety Notes**

1. Direct supervision is required during all aspects of this activity to ensure that safety behaviors are followed and enforced.

2. Make sure that any items dropped on the floor or ground are picked up immediately after working with them—a slip/trip fall hazard.

3. Use caution in working around electric lamps. They get hot and can burn skin. They also can shatter if exposed to liquid like water.

4. Electric lamps should only be plugged into GFI-protected receptacles to help prevent shock. Keep electrical wires away from water sources too.

5. Use caution in working with sharp objects such as wires; they can cut or scratch skin.

6. Follow the teacher's instructions for returning materials after completing the activity.

7. Wash your hands with soap and water after completing the activity.

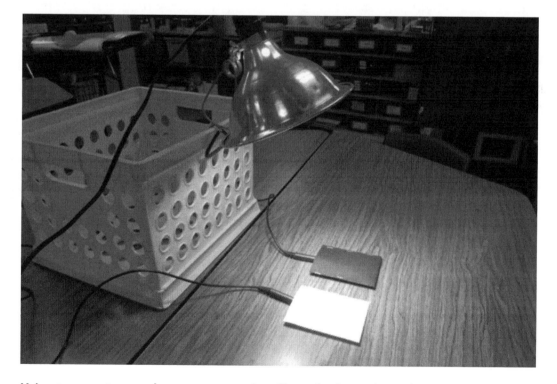

**Using temperature probes to measure the effect of color on heat absorption.**

| Time in Minutes | Temperature in Degrees Celsius | | | | | |
|---|---|---|---|---|---|---|
| | Cotton | Wool | Paper Towel | Foil | Bubble Wrap | No Cover |
| 0 | 4.5 | 4.6 | 4.7 | 4.5 | 4.0 | 5.8 |
| 30 | 8.0 | 7.6 | 7.7 | 7.0 | 7.9 | 10.2 |
| 60 | 10.9 | 10.2 | 10.1 | 9.3 | 10.8 | 13.4 |
| 90 | 13.1 | 12.3 | 12.1 | 11.2 | 13.1 | 15.6 |
| 120 | 14.9 | 14.1 | 13.8 | 12.8 | 15.0 | 17.2 |
| 150 | 16.3 | 15.6 | 15.1 | 14.2 | 16.3 | 18.4 |
| 180 | 17.5 | 16.7 | 16.2 | 15.3 | 17.4 | 19.2 |

**TABLE 8.2. Temperature of Water Insulated in Various Materials Over Time**

I then asked, "Which of the coverings do you think kept the water cooler the longest? Why do you think this?" I asked the students to write their ideas in their science notebooks after they had taken the time they needed to think about the questions on their own. I gave them some time to think and then walked over to my student Jimmy and asked him what he thought. I knew that he would not write in his notebook, so I had him verbally articulate his ideas and I wrote down his ideas.

After everyone was done writing, I provided students with the data I had collected and asked them how we might use them to determine which material was the best insulator. Right away students called out, "We can use HLPA!" HLPA is a strategy that they had been taught to use when looking at data and graphs. Students examine high points, low points, patterns, and anomalies as a way to review and make sense of the data as a whole. HLPA strategy is as follows:

- *H*—High values
- *L*—Low values
- *P*—Patterns
- *A*—Anomalies

**UDL Connection**
*Principle II: Action and Expression*
*Guideline 5: Expression and Communication*
*Checkpoint 5.1: Use multiple media for communication*

Once the students had conducted an initial scan, I asked them to revisit the question about which type of material is the most effective in preventing heat transfer. In small groups, the students worked together to examine the data, write down a claim and the evidence that supports that claim, and explain their reasoning. The claim-evidence-reasoning (C-E-R) framework is one that I've used many times to scaffold students' sense-making (McNeill and Martin 2011). The ideas that students recorded in their notebooks now became a tool for us the next day as we moved to the Explain phase of the cycle.

> **Teaching Tip:** While in many lessons in this unit I involved students in planning and conducting the investigation, I set up these two investigations and focused students on interpreting the data. This is a scaffold for helping them engage in data analysis as they evaluate their designs later in the lesson.

## Explain Phase

Sense-making and engaging in argumentation from evidence can be challenging for students—and for teachers. My role in this phase of the lesson was to think and reason along with students, not to think for them. Below is a re-creation of our discussion:

Teacher: *Let's talk about the ideas about the insulator data that you came up with yesterday. What did you notice about this data set?*

Scott: *The bottle of water with no cover had the highest temperature at the end, so it was the warmest.*

Addie: *The bottles with the cotton sock and the bubble wrap had almost the same temperature at the end. That means they were the second warmest.*

James: *We thought that the foil was the best because it had the lowest temperature of all the bottles at the end of three hours.*

Teacher: *All of you are using the last temperature reading as evidence to support your claims?*

Molly: *Yeah, but none of the bottles had the same beginning temperature. The one without a cover was the highest temperature to begin with.*

Teacher: *Does that make a difference?*

Lee: *No, I think it is the ending temperature that counts. It doesn't matter what the beginning temperature is.*

Molly: *I think the beginning temperature does matter!*

Teacher: *Why do you think that?*

Molly: *Because if the beginning temperature is different and the ending temperature is different, then we don't really know which of them changed more just from the ending temperature.*

Teacher: *Can anyone else express Molly's idea in their own words?*

Max: *It's like you don't know how much you have grown if you only know how tall you are. You need the first temperature too.*

Teacher: *Taking into account the idea that Molly and Max just shared, what should we use as evidence for a material being a better or worse insulator? Talk about that in your table groups for a minute.*

After talking with their groups, the students reached a consensus that they should examine the *change in temperature* that each bottle underwent.

Before they begin calculating, I asked, "So, if one of the bottles has the greatest change in temperature, does that mean it is the best or worst insulator in the group we tested?" The class immediately responded with a chorus of "Best!" and "Worst!" I asked for volunteers to share their reasoning:

Lee: *If it changed more, then the material was better.*

Teacher: *Better at what? (I probe.)*

Lee: *At heating up … Um. No! I changed my mind! Worst!*

Teacher: *Can you tell me why you changed your mind?*

Lee: *We want the nest to be cool.*

Teacher: *Who can add to that?*

Scott: *The one that stayed cool was protected from the heat.*

Sonia: *The best insulator will slow the heat the most.*

The groups got to work subtracting the initial temperature from the final temperature and comparing the results. I then asked each group to revisit their initial claim, evidence, and reasoning and provide a revision on their whiteboard to share with the class. We then held a gallery walk (Keeley 2014), in which students circulated, read each group's claims, and convened to discuss any areas of disagreement. Though worded slightly differently, all groups agreed that the foil was the most effective insulator (claim) because it had the lowest change in temperature among the materials (evidence), meaning that it prevented the transfer of heat energy (reasoning).

> **Teaching Tip:** Rather than having the students share orally, and thus requiring students to access information through listening, a gallery walk allows students to move at their own pace as they read the information written on each group's board to themselves, or with the support of a peer.

We repeated the same process with the data we collected from our heat pockets investigation (see Table 8.3, p. 177), with students determining that the black object absorbed more heat than the white one (claim), because the temperature increase was greater (evidence). The reasoning for this one was a bit more challenging for students, so I reminded them of our earlier learning and asked them what temperature indicated—the amount of heat energy. Because of this, we could say that a larger temperature increase was an indication of more energy being transferred from the lamp to the pocket (see Figure 8.2, p. 176).

## Extend Phase

It was time to apply what the students had learned about heat to solve the penguin problem! I used a problem-based activity, as it is a relevant and novel problem that engages all of my students equally. Because the students participated in a full engineering design cycle, this phase was pretty extensive (pun intended!). As an introduction to engineering design, we watched "The Engineering Process: Crash Course Kids #12.2" on YouTube (*www.youtube.com/watch?v=fxJWin195kU*). The following steps are described:

1. Define the problem.
2. Do research.
3. Develop a possible solution.
4. Design your solution.
5. Build a prototype.
6. Test it.
7. Evaluate your solution.

I provided a refresher of the problem and constraints, along with criteria for success in the form of a "client card" (Capobianco, Nyquist, and Tyrie 2013). I told the class the following:

*African penguins are in need of a nest cover to protect them from the heat. The prototype you design should have a place for a penguin to enter, and be no larger than 5 inches by 5 inches by 5 inches. Materials should be easy to obtain and low cost, and they should be effective in protecting your penguin from overheating.*

The "penguin" in this scenario was made of ice! The students' prototypes would be tested using an ice-cube penguin that would be weighed before being placed in the nest. A lamp (Sun) would be turned on to shine on the nest for 20 minutes. At the end of 20 minutes, the ice penguin would be weighed again. Students would calculate the weight loss. "Overheating" was defined as an ice penguin losing more than half its mass to melting.

**Teaching Tip:** I chose the ice penguin as an alternative to the temperature probes, as it provided students with a more concrete visualization of the transfer of heat occurring in the nests. However, the temperature probes could certainly be used in this scenario.

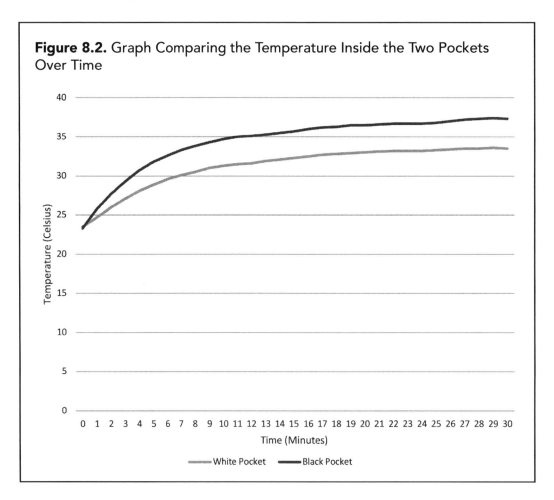

**Figure 8.2.** Graph Comparing the Temperature Inside the Two Pockets Over Time

| Time in Minutes | Temperature in Degrees Celsius | |
|---|---|---|
| | White Object | Black Object |
| 0 | 23.5 | 23.3 |
| 1 | 24.7 | 25.8 |
| 2 | 26 | 27.7 |
| 3 | 27.1 | 29.3 |
| 4 | 28.1 | 30.7 |
| 5 | 28.9 | 31.8 |
| 6 | 29.6 | 32.6 |
| 7 | 30.1 | 33.3 |
| 8 | 30.5 | 33.8 |
| 9 | 31 | 34.3 |
| 10 | 31.3 | 34.7 |
| 11 | 31.5 | 35 |
| 12 | 31.6 | 35.1 |
| 13 | 31.9 | 35.3 |
| 14 | 32.1 | 35.5 |
| 15 | 32.3 | 35.7 |
| 16 | 32.5 | 36 |
| 17 | 32.7 | 36.2 |
| 18 | 32.8 | 36.3 |
| 19 | 32.9 | 36.5 |
| 20 | 33 | 36.5 |
| 21 | 33.1 | 36.6 |
| 22 | 33.2 | 36.7 |
| 23 | 33.2 | 36.7 |
| 24 | 33.2 | 36.7 |
| 25 | 33.3 | 36.8 |
| 26 | 33.4 | 37 |
| 27 | 33.5 | 37.2 |
| 28 | 33.5 | 37.3 |
| 29 | 33.6 | 37.4 |
| 30 | 33.5 | 37.3 |

**TABLE 8.3. Data Collected From the Heat Pockets Investigation**

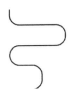

**UDL Connection**

*Principle III: Engagement*
*Guideline 7: Recruiting Interest*
*Checkpoint 7.2: Optimize relevance, value, and authenticity*

In the previous lessons, students had already conducted research that could help inform their designs, so that is where we began. I intentionally paired up students to work on the design of their nest cover. My goal was to put students together in a way that would use their strengths to support their weaknesses, both academically and behaviorally, as both were critical in this project.

**UDL Connection**

*Principle III: Engagement*
*Guideline 7: Recruiting Interest*
*Checkpoint 7.3: Minimize threats and distractions*

The teams began by discussing how a nest box might solve the problem and by developing a plan and sketch for their prototype. They needed to decide not only how they would build it, but also what materials they would use. I am a bit of a pack rat, and I save all kinds of items that can be used for science. Sometimes people give me items or materials they think I might be able to use. The availability of items varies each year. The following is a representative list of materials that I have made available for students to use in building their prototypes: polystyrene foam; cardboard; fabrics such as fleece, cotton, and denim; foil; bubble wrap or bubble-wrap mailing envelopes; string; yogurt or sour cream containers; craft foam; felt; craft sticks; and masking tape. Each material had a cost attached to it by unit. The unit might be a squared inch, inch, or foot. Materials that are scarce were priced accordingly. The cost of the materials used in the final design was calculated and used as the single unit cost per nest box for evaluation purposes.

**UDL Connection**

*Principle II: Action and Expression*
*Guideline 6: Executive Functions*
*Checkpoint 6.2: Support planning and strategy development*

The students needed to maintain a design log. Before they got any materials, they had to submit an initial design drawing and materials list. I had set up a testing station with a couple of lamps and temperature probes that they could use to test the insulation properties of the materials they had selected, and they would conduct a pilot test of their design to troubleshoot any potential problems.

The students often identified changes they wanted to make to their designs; however, they could not get any additional materials without showing their redesign ideas with a rationale behind the change. Because some students struggled with maintaining a log that might be writing intensive, I gave the option of using a camera and Google Drawing or an iPad with a notation app to document their design changes.

> **UDL Connection**
> *Principle II: Action and Expression*
> *Guideline 5: Expression and Communication*
> *Checkpoint 5.1: Use multiple media for communication*

## Evaluate Phase

Once the students had built their final prototypes, we had a formal testing day (see the photos below and on p. 180). Before testing, students measured the dimensions of their nest cover to determine if it fit within those criteria. Then they verified the measurements by trading their nest cover with another pair to measure the dimensions. They also had to have a complete list of the materials they used and the amounts of each. While the test was being conducted, the

**Student-designed prototype nest cover.**

**Tasting a variety of materials and their performance as insulators.**

partners were to calculate the cost of their prototype using their materials list and the cost sheet.

On test day, all penguin ice cubes were weighed using small digital balances that measured the weight in grams to the nearest tenth. I had four boxes for testing and eight lamps. This accommodated a total of 12 to 16 nest covers. Nest covers were to stay under the lights for 20 minutes (see the photo below). As

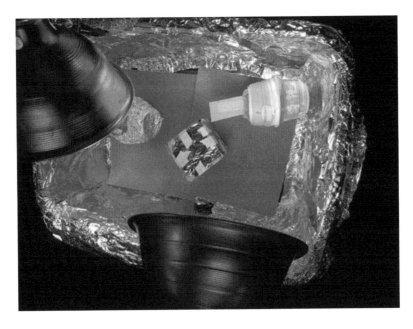

**Testing the effectiveness of students' prototype designs.**

much as possible, I attempted to ensure that all nest covers got an equal amount of light.

After testing, the water was poured off of the penguins and they were weighed again. Students then calculated the amount of weight loss in grams.

Each pair then prepared an evaluation form for their design that indicated (1) the total cost for assembly, (2) whether it met the constraints (e.g., appropriate size), and (3) the amount of weight loss experienced by the ice penguin during testing. These were placed next to the prototypes on display for all class members to review in a gallery walk format.

As students circulated around looking at different designs, they realized that some designs worked well, but cost more; others worked slightly less well, but were less expensive; and so on. I drew on those observations in our whole-class discussion to help students understand the trade-offs inherent in their designs.

We prepared a class chart that aggregated our results. The first column indicated by a ✓ or an ✗ whether the group met the constraints of the dimensions for their prototype. The second column listed the total cost per unit of their design, and the third column listed the amount of mass lost (grams) in the ice penguin during testing. As had become a familiar routine, students use the HLPA strategy to identify those designs that performed best. Typically, as the students noticed in the gallery walk, there was no clear "winner"—our top performer as an insulator was among the most costly, prompting questions about how they might reduce costs. Students also noticed patterns in terms of which materials were used by the designs in which the ice penguins lost the least amount of mass, prompting them to think about how they might use those to improve their own designs.

Each pair submitted a self-evaluation of their design—what they thought worked well, and what they would improve and why. I used this, along with a review of their design logs, to evaluate their engagement in the engineering design process and potential areas in which they might need support in our next engineering task.

I emphasized that scientists and engineers will often go through several different iterations and prototypes before settling on one design, and even then they will still seek to improve it as new problems are encountered or technologies improve (bringing up the newest iPhone is usually an example that resonates with them). I then shared a news article with the students that detailed problems that conservationists had experienced implementing nest boxes for African penguins. In this article, the author wrote,

> These attempts have seen very limited success, and in some attempts they inadvertently introduced new challenges for the penguins. In many cases, the penguins chose not to use the nesting

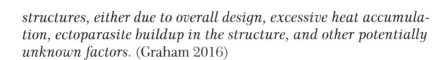

*structures, either due to overall design, excessive heat accumulation, ectoparasite buildup in the structure, and other potentially unknown factors. (Graham 2016)*

We discussed that while our tests might work well for our "ice penguins," they don't indicate how these designs might perform in the real world. The activities in this phase of the lesson address all three dimensions of the *NGSS*. However, because we had conducted these activities in pairs, or collaboratively, throughout the discussion, this doesn't necessarily provide me with a summative assessment of individual student learning. Nonetheless, it provided me with valuable information as to what my students could do well, and where they might need support to prepare for the individual engineering design project that would occur in our next unit of study.

## Unpacking UDL: Barriers and Solutions

Table 8.4 summarizes the UDL principles, guidelines, and checkpoints that I applied when designing the activities for each phase of the learning cycle lesson to meet the *general* needs of the learners in my classroom. Below is an example of how I identified barriers and strategized solutions to meet the *specific* needs of one of the learners in my classroom.

### *Learner Profile: Jimmy*

"Jimmy" had difficulty working with others and was a reluctant writer. He tended to want total control of a project or would not participate at all. When working with partners during the Extend phase, I purposely partnered Jimmy with Jacey, because she has patiently worked with him in the past and was able to cajole him into working collaboratively.

In the Explore part of the lesson, students were asked to think about the materials and write their prediction about which of the insulators they thought would be most effective. I gave the class some time to think and then walked over to Jimmy and asked him about his prediction. I knew that he would not write in his notebook, and this was a way for him to articulate his ideas. Because my purpose in having students write their prediction was to get students to think about the materials as insulators, I was not concerned about whether Jimmy wrote this down. He expressed his ideas orally, and I was able to scribe for him to refer back to them.

In the Extend phase of the lesson, I provided various media options (e.g., a camera, an iPad, writing in the notebook) as a way to keep a design log.

| TABLE 8.4. UDL Connections | |
|---|---|
| **Connecting to the Principles of Universal Design for Learning** | |
| ***Principle I. Representation*** | |
| *Guideline 3: Comprehension* | |
| Checkpoint 3.1. Activate or supply background knowledge | Some of my students struggle to connect what they have previously learned to the current task at hand. To help with this lesson, I deliberately prompt my learners via questions to recall key concepts from prior lessons. Prior to the lesson, the students were taught a strategy to help them engage in graphs and data. |
| ***Principle II. Action and Expression*** | |
| *Guideline 5: Expression and Communication* | |
| Checkpoint 5.1. Use multiple media for communication | During the Extend and Evaluate phases of the lesson, I provide different approaches and media to communicate ideas as a way to ensure that my struggling writer has a way to communicate and record his ideas. |
| *Guideline 6: Executive Functions* | |
| Checkpoint 6.2. Support planning and strategy development | During the Extend phase, the students are required to develop a plan for their prototype as a support for solving the problem scenario. |
| ***Principle III. Engagement*** | |
| *Guideline 7: Recruiting Interest* | |
| Checkpoint 7.2. Optimize relevance, value, and authenticity | This lesson involves a novel, complex, and authentic problem-based situation to engage all my learners equally. |
| Checkpoint 7.3. Minimize threats and distractions | Strategic partner arrangements used during the Extend phase provide a way to create a supportive classroom climate for carrying out tasks. |

> ## Questions to Consider
>
> ➤ To what extent did the activities in this lesson align with the purpose and intent of each phase of the 5E Learning Cycle? Could you envision other activities that would be appropriate for each phase?
>
> ➤ Were you able to follow the sequence of activities and the ideas that students developed in the lesson? How did the storyline of the lesson progress? Do you agree with the teacher's choice to introduce the Engage phase activities for this final learning cycle at the beginning of her unit, and then return to the cycle later?
>
> ➤ In what ways was the teacher able to assess students during each phase of the lesson? How did this inform future instruction?
>
> ➤ In what ways did the solutions that the teacher identified meet the needs of the specific students spotlighted in this vignette? In what ways did they benefit all students? Could you think of other solutions you might use for your own students?

## References

Braun, D. M. 2010. African penguin declared endangered. *National Geographic*, June 2. *https://blog.nationalgeographic.org/2010/06/02/african-penguin-declared-endangered*.

Capobianco, B. M., C. Nyquist, and N. Tyrie. 2013. Shedding light on engineering design. *Science and Children* 50 (5): 58–64.

Graham, K. 2016. African penguin artificial nest development workshop in Cape Town, South Africa. *www.aza.org/aza-safe-stories/posts/african-penguin-artificial-nest-development-workshop-in-cape-town-south-africa*.

Keeley, P. 2014. *Science formative assessment: 75 practical strategies for linking assessment, instruction, and learning* (Volume 2). Thousand Oaks, CA: Corwin Press.

McNeill, K. L., and D. M. Martin 2011. Claims, evidence, and reasoning: Demystifying data during a unit on simple machines. *Science and Children* 48 (8): 52–56.

NGSS Lead States. 2013. *Next Generation Science Standards: For states, by states.* Washington, DC: National Academies Press. *www.nextgenscience.org/next-generation-science-standards*.

Sheerer, K., and C. Schnittka. 2012. Save the Boulders Beach penguins. *Science and Children* 49 (7): 50–55.

# CHAPTER 9

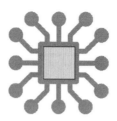

# Making Connections
# With Circuits

**Nicole Burks**

This chapter introduces a lesson in which students learn about conductors and insulators and then use this learning to better understand how their science equipment for circuits works *and* how an electrical toy functions. The lesson comes after an initial learning cycle in which the students have formed a complete circuit to light a bulb using only a battery and a wire, and it is spread out over several days. The lesson-level phenomenon ("How does this toy work?") provides a familiar and motivating context and prepares students to figure out an overarching unit phenomenon related to energy conservation—how a car can run without gas. The conceptual flow of the lesson prepares students in identifying the parts of the circuit and how they work together, so that they can begin to consider what is going on within the system in terms of energy in later lessons. Figure 9.1 (p. 186) displays a conceptual storyline of the lesson.

---

**Figure 9.1.** Conceptual Storyline of the Lesson

**Engage**
Brainstorming solutions to a problem

**Key idea:** There may be more than one solution to a problem

*Linking question:* What can we learn to help us test our ideas?

**Explore**
Testing materials to see which complete a circuit

**Key idea:** Some materials work and others don't

*Linking question:* What kinds of materials work and what kinds do not?

# How Can We Connect With Circuits?

**Explain**
Defining "conductor" and "insulator"

**Key idea:** Metals are conductors and complete a circuit; non-metals are insulators and do not complete a circuit

**Evaluate**
Designing a game using conductors and insulators

**Key idea:** Energy flows to the bulb when the circuit is completed

*Linking question:* What can we design using conductors and insulators?

**Extend**
Examining sockets and battery holders

**Key idea:** Conductors make contact with the battery and bulb in sockets and holders

*Linking question:* How do we use conductors and insulators?

## LESSON VIGNETTE

Growing up, I noticed that my father only wore his wedding ring on fancy occasions. It was a detail that I always noticed, but never reflected upon. I never questioned it, until I was in fifth grade learning about electricity. My father is an electrician. His wedding ring is metal. As my teacher was explaining the difference between insulators and conductors, I experienced an intense "Aha!" moment. The switch flipped, the circuit was complete, and the light bulb went on. I will always remember the difference between a conductor and an insulator because of this personal connection. Relevant learning experiences are critical to develop student understanding. They are extremely important in science when students are learning about new or abstract ideas.

Summers are rarely a vacation for educators. We spend our "off" time attending professional development, teaching summer school, or reading curricula and books for the school year. In the summer of 2015, I participated in the Quality Elementary Science Teaching (QuEST) program. It was a two-week program that allowed us to explore, question, and learn about teaching about electrical circuits and energy within the classroom. The first week was filled with purposeful learning experiences that we could bring into our classroom for the upcoming school year. I found myself walking out of the two-week program excited about implementing what I had learned with my incoming fourth graders.

I left the workshop excited about the possibility of having my students experience the same "Aha!" moment that I did in that fifth-grade classroom. Rather than telling my students about conductors and insulators, I wanted to teach them in a way that allowed them to experience for themselves and develop an understanding of different materials and whether they complete a circuit. I wanted them to be able to relate what they were learning to their lives outside of school. Most importantly, I wanted to know what they thought was going on when the circuit was complete, so that I could build on their initial ideas about energy in subsequent lessons.

See Table 9.1 (p. 188) for alignment of this lesson to the *Next Generation Science Standards* (*NGSS*; NGSS Lead States 2013).

| TABLE 9.1. *NGSS* Alignment | |
|---|---|
| **Connecting to the *NGSS*—Standard 4-PS3: Energy** | |

*www.nextgenscience.org/dci-arrangement/4-ps3-energy*

- The chart below makes one set of connections between the instruction outlined in this chapter and the *NGSS*.
- The materials, lessons, and activities outlined are just one step toward reaching the performance expectation listed below.

**Performance Expectation 4-PS3-4.** Apply scientific ideas to design, test, and refine a device that converts energy from one form to another.

| Dimensions | Classroom Connections |
|---|---|
| **Science and Engineering Practices** | |
| *Constructing Explanations and Designing Solutions*<br><br>Apply scientific ideas to solve design problems. | Students use scientific concepts and learn about simple circuits to design and build an electronic game. Going through this design process for a simple system prepares them for designing a more complex system. |
| **Disciplinary Core Ideas** | |
| *PS3.B: Conservation of Energy and Energy Transfer*<br><br>Energy can also be transferred from place to place by electric currents, which can then be used locally to produce motion, sound, heat, or light. | Students build an electrical device that produces light. In subsequent lessons, they will add to the circuit/system to incorporate motion and to use a solar cell instead of a DC battery. |
| **Crosscutting Concepts** | |
| *Energy and Matter*<br><br>Energy can be transferred in various ways and between objects. | While students are transferring energy (potential energy to electrical energy to light energy) in their electrical device, they do not yet address this concept. Later in the unit, they will use an energy lens to analyze the device they built. |
| *Systems and System Models*<br><br>A system can be described in terms of its components and their interactions. | Students describe their device in terms of its parts and how they work together. |

## Engage Phase (Day 1)

I began the lesson by strategically placing students into small groups with peers who had complementary strengths or weaknesses. Within the small groups, students were asked to solve the following problem:

*You are on a camping trip and realize that you only have one bat-
tery with you to power your flashlight, which requires two. How
might you solve this problem? (Hint: You packed your sandwich
in aluminum foil and your lunch in a brown paper bag.) Come up
with a plan. Sketch your possible solution.*

In addition to providing a paper copy to students, I projected the problem on the
board and read it aloud to the class.

**UDL Connection**

*Principle III: Engagement*
*Guideline 8: Sustaining Effort and Persistence*
*Checkpoint 8.3: Foster collaboration and community*

**UDL Connection**

*Principle I: Representation*
*Guideline 1: Perception*
*Checkpoint 1.3: Offer alternatives for visual information*

**UDL Connection**

*Principle II: Action and Expression*
*Guideline 6: Executive Functions*
*Checkpoint 6.3: Facilitate managing information and resources*

During this part of the lesson, I made it a point to not get involved in my students' conversations or make suggestions. In this phase, it was important for the
students to be engaged by the problem and to begin applying their prior knowledge. It was their turn to think and question, not mine. As the groups discussed
what they would do if this scenario happened to them, I circled around the room.
I spent most of my time listening to my students. I wanted to hear what prior
knowledge they were applying to this situation:

• Were they using vocabulary they had learned about in the previous learn-
  ing cycle?
• Were they thinking about previous personal experiences?
• Were they building off one another's thinking?
• Were they using the inference clues within the scenario?

I encouraged the students to put their thinking on paper using either images
or words or both. Giving students a choice allows them to represent their thinking
in a way that students can best communicate what they understand.

**UDL Connection**
*Principle II. Action and Expression*
*Guideline 5: Expression and Communication*
*Checkpoint 5.1: Use multiple media for communication*

Once the students were done working within their small groups, I brought the entire class together to discuss and chart their predictions for how to solve the problem. We met in a familiar routine for our classroom, a dinner table conversation. During this type of conversation, no hands need to be raised. All ideas can be shared and should be respected. Students build off one another's thoughts, they don't just blurt out to be heard. They ask questions about what others are saying when they don't understand.

The students sat in a circle on the floor, rather than around an actual table, and they brought their notebooks with them as a tool to support their talk. I was amazed at the ideas and suggestions they were giving. Students were beginning to develop reasonable predictions and were using concepts from the previous lesson when sharing their ideas. Some suggested making the foil into a wire and using it to connect the end of the battery to the bulb. Others suggested forming the foil into the shape of a battery and putting it into the flashlight in place of the second battery. Still others thought removing the bulb from the flashlight and connecting it to the foil would be necessary.

Listening to my students' predictions and reasoning helped me plan accordingly for the next day.

## Explore Phase (Day 2)

I wanted my students to start a new day with the same type of passion that they ended with the day before. It was important to remind them of their predictions, yet take the exploration to the next step. I planned to reach that next step by asking my students questions that they hear fluidly throughout the day: "Why?" "How do you know?" "Can you provide evidence to support your thinking?"

With the same partners from yesterday, students built a circuit tester by connecting a battery, bulb, and wires so that there is a gap between two wires that—when touched—complete the circuit and light the bulb. The students were then given a bag of various items to place within the gap. Their goal was to find out whether the circuit was completed by different objects, as evidenced by the bulb lighting. I was very strategic about the items I chose to place inside the bag. I wanted some obvious contenders (metal) but also chose some items that would push students' thinking further (i.e., a pipe cleaner, clothespin, penny, and plastic fork with a metallic-looking finish). See the "Materials and Safety Notes" box.

## Materials and Safety Notes

**Materials**

| | |
|---|---|
| Batteries | Clothespins |
| Bulbs | Pennies |
| Wires | Plastic forks with metallic-looking finish |
| Bags | Battery holders |
| Metals | Bulb holders |
| Pipe cleaners | Battery clips |

**Safety Notes**

1. All involved must wear indirectly vented chemical splash goggles or safety glasses with side shields during all phases of these electricity inquiry activities (setup, hands-on investigation, and take-down).
2. Direct supervision is required during all aspects of this activity to ensure that safety behaviors are followed and enforced.
3. Make sure that any items dropped on the floor or ground are picked up immediately after working with them—a slip/trip fall hazard.
4. Use caution in working with sharp objects such as wires; they can cut or scratch skin.
5. Use caution when working with bulbs. Glass is fragile; it can break and cut skin.
6. Never stick wires into electrical receptacles—this can burn or shock the user.
7. Don't keep circuits connected too long. Wires can heat up and cause burns or even a fire.
8. Follow the teacher's instructions for returning materials after completing the activity.
9. Wash your hands with soap and water after completing the activity.

Before the exploration began, we took some time to discuss how they would organize their findings and data within their science notebooks. Again, I wanted the students to be able to represent their observations in a way that made sense to them. People learn and share their results in different ways, and that is something that I try to value in my classroom. Some students opted to use graphic organizers (their choice of two-column chart or Venn diagram) as a tool to organize their notebooks and allow them to easily locate information and make sense of it.

**UDL Connection**

*Principle II: Action and Expression*
*Guideline 6: Executive Functions*
*Checkpoint 6.3: Facilitate managing information and resources*

Once we were prepared, the exploration began, and I circled around the classroom asking students guiding questions. Not once did I give them a definition or a scientific name, though at times students brought up terms with which others may not be familiar. For example, my student James asked, "Aren't all

metals conductors?" I neither confirmed nor denied, but rather asked for elaboration and evidence of his claim.

I asked the following: "What do you mean by conductors?" "How do you know?" "Can you show me an example of what you are explaining?" At this moment in the exploration, it was important for my students to start to develop their own understanding of what was happening when they completed the circuit with the random items. I did not want to provide them with the scientific name until they could explain, in their own words, what was happening during this exploration. This helped them associate their own experience and meaning with any new terms I introduced.

As I made my rounds around the classroom, I noticed Jerry deep in conversation with Sally about the pipe cleaner. Jerry believed that he could make the bulb light when the pipe cleaner was placed within the gap. However, Sally disagreed; she claimed that this would not work because of the fuzz on the outside. She said, "The cotton ball didn't complete the circuit, so why would the pipe cleaner?" Sally was more focused on the fuzziness of the material, rather than considering what was inside the pipe cleaner. I may not have been involved in the discussion at this point, but I was preparing for our conversation to come in the Explain phase.

I let their conversation continue and made a note to focus our discussion on the items made of more than one type of material. I stood back and noticed that my student Eddie was going to grab his pencil to test in the circuit. Then Mary instantly wondered if her iPad cover would work. The students were inquiring further on their own. Everyone was in a different place, yet all were challenging their prior knowledge and questioning other possibilities. For students who might lose focus or struggle to recall what to do, I referred them to a bulleted list on the board that I had created to help with managing the step-by-step process for testing an item with their circuit. I also projected an image with my document camera on how the circuit tester is set up.

**UDL Connection**

*Principle II: Action and Expression*
*Guideline 6: Executive Functions*
*Checkpoint 6.2: Support planning and strategy development*

## Explain Phase (Day 2, Continued)

Although questions, connections, and conversations were still happening at their table groups, I could tell we had some key ideas to discuss, so I signaled to gain the attention of my students. As the students made their way to the carpet for a "dinner conversation," I asked them to bring their science notebooks filled with their data and observations. The notebooks would not serve the purpose of a seating place or a sketch board, but as a tool that would support their thinking with data from experience. (We were scientists in Room 210, so we argued using evidence.)

**Teaching Tip:** If students opt to use different kinds of graphic organizers for their data, this can also be a time to evaluate the strengths and weaknesses of those choices and to help students develop skills for organizing data in appropriate ways.

Before we began, I reminded students that I wanted them to think and hear and consider all the evidence before they changed their thinking. I started by asking, "What do all of the objects that completed the circuit have in common?" During our conversation, the students reached a consensus based on their evidence—all items that completed the circuit were metal or had metal parts. The non-metal items did not allow the bulb to light. I finally gave my students a name for that group of objects—when it was necessary to describe what they had observed firsthand.

In their science notebooks, students defined the scientific terms *insulator* and *conductor*:

- **Conductor**—an object that can complete the circuit, allowing the electrical current to flow through the circuit and to make the bulb light. Metals are conductors.
- **Insulator**—an object that does not complete the circuit. Plastic, glass, and paper are conductors.

The students drew a box around key vocabulary terms and used highlighters to make those terms stand out in their notebooks, so they could find them easily when needed.

**Teaching Tip:** Students can also create a glossary in the back of their notebooks for easy reference and to find all vocabulary in one place.

I started thinking to myself, "Do I want to end our conversation? Where do we need to go next? What did our facilitators do to challenge us during QuEST?" I was not going to let this "dinner conversation" end. We needed to take our thoughts further:

Teacher: *What other objects do you think would be conductors or insulators?*

Mary: *I tried out my iPad case, and it did not complete the circuit.*

Teacher: *Why do you think that?*

Mary: *Well, our iPads cases are rubber, and we found that rubber is an insulator today.*

I encouraged the students to consider other objects (metal or non-metal) that they encountered every day:

Teacher: *Why might it be important to understand what objects are insulators and conductors in our everyday lives?*

Jerry: *Don't electricians need to think about what materials they are working with so they don't get electrocuted?*

Sam: *We have plastic things on our plugs so my baby brother doesn't stick things into them.*

The students were using the concepts to make sense of their everyday experiences. New connections were being made. The light bulbs were going off in my students' minds. (Puns intended!) My goal for this discussion was accomplished, yet, from my QuEST professional development experience, I was aware that my job was far from done. To assess what my students were thinking individually, I asked them to work on the following task as an exit slip:

*Using what you have learned, revisit your answer to the camping scenario. Decide whether or not you still agree with your original thought, and explain any changes you would make and why.*

Providing the sentence starters "I used to think …" and "Now I know …" (Keeley 2008) gave the students who are reluctant writers a starting point for expressing their ideas.

**UDL Connection**
*Principle II: Action and Expression*
*Guideline 5: Expression and Communication*
*Checkpoint 5.2: Use multiple tools for construction and composition*

As I reviewed the students' work to prepare for the next phase of the lesson, I considered the following:

- Were students using the vocabulary and concepts from the first learning cycle to describe their solution (identifying parts and how they are connected)?
- Were students using the scientific terms *insulators* and *conductors* from this lesson in their explanation?
- Were students using evidence to support their reasoning?
- What were students' explanations for why their solution worked? How, if at all, did they incorporate energy (an upcoming concept) into their explanations?

Reviewing the students' work helped me decide who to check in with and who might need additional support or challenge in the next phase. The last question in particular helped me understand the prior knowledge that students had about what was happening *inside* the circuit and whether they related that to energy, which we would explore in greater depth in the next learning cycle.

## Extend Phase (Day 3)

To (re)activate students' prior knowledge, I began the next day of this learning cycle by showcasing two example explanations that I had generated from the students' exit slips from the previous lesson. I was strategic with the examples I chose to share, pairing two different solutions (anonymously) for students to compare and evaluate. I made sure to include common errors in one solution and provide a model for a good explanation in the other. This helped the students self-evaluate and identify any errors in their thinking, and it also helped them understand what makes a good explanation and how they could improve.

Next, I placed three new items on each group's table: a battery holder, bulb holders, and battery clips. I asked the students to examine each and decide which parts of the holders and clips are insulators and conductors. Then, students constructed a complete circuit using these new materials. Up until this point we had been using just batteries, bulbs, and wires. We had focused on the important contact points made between these items when the circuit was complete, so the students were able to relate this to the contact made between the socket and parts of the bulb, and the holder and ends of the battery.

Once the groups had successfully assembled the circuit, I asked them to share their ideas using an app called Screen Chomp. (Similar apps such as Explain Everything could serve this purpose as well.) The students took a picture of their completed circuit and recorded the group explaining how it works.

Two sample videos made by my students can be found online:

- *https://youtu.be/SSBL3AoIUhE*
- *https://youtu.be/U6HiB5dmLqw*

I took the time to review the videos created by each group for accuracy. These videos provided evidence that the students understood how the different components formed a complete circuit, but they were confusing the flow of the electrical current versus the flow of energy in the circuit. I made a note to attend to students' use of the word "energy" so that I could build from that in the next learning cycle.

## Evaluate Phase (Days 4 and 5)

I began the final phase of our learning cycle by showing a video clip featuring a popular electrical game that tests dexterity. Several versions of this game exist, including "Don't Buzz the Wire!" and "Beat the Buzz," and videos showing the game in action can be found on YouTube. After the video, I asked students to talk in partners about how they thought the game worked and how it related to what they had been learning. Several students also brought up similar games they had played such as the "Operation!" game.

For a final task, I presented students with the challenge of working with a partner to create their own version of the game that lights up, rather than buzzes, using materials found in our classroom. While it seems quite simply to be designing something that emulates an existing object, it is not as straightforward as you'd expect! We participated in a similar task in the QuEST program, and I found it both challenging and motivating as it was an authentic task that allowed for active participation, exploration, and experimentation. I remember collaborating with others and resolving disagreements about how to proceed, getting stuck, failing, problem-solving, and finally succeeding. I anticipated that I would encounter similar responses among my own students, many of whom thrived and persisted in tasks that were meaningful to them.

In tackling the design process first in this manner, students would be building the skills to enable them to be successful in more challenging design tasks later in the unit. The students do not meet NGSS 4-PS3-4, "Apply scientific ideas to design, test, and refine a device that converts energy from one form to another," but they take the first step to applying what they've learned about circuits to design a device.

> **Teaching Tip:** This particular game can have many different designs and allows students to see multiple ways to approach building something. However, you could use other circuit-based games or can provide students with the opportunity to design their own.

We began our engineering design process by brainstorming possible designs. Students, with the guidance of a graphic organizer, discussed their plans before beginning to build. I asked questions to help them think through their plans, such as "What prior experiences helped you in making your plan?" "What materials did you use? Why?" "How do the parts connect?" and "When is the circuit complete?"

As I walked around the room, I was amazed at what my students were doing and how engaged they were in the task. Mary was explaining to her partner why she wanted to wrap the guiding metal tool in cloth or rubber bands. "I think it is important to have an insulator as the handle so the person doesn't touch the metal," she explained. There were disagreements and questions, but as a group they agreed that she should incorporate that into the design.

**UDL Connection**

*Principle II: Action and Expression*
*Guideline 6: Executive Functions*
*Checkpoint 6.3: Facilitate managing information and resources*

As plans were finalized, the students began to collect the materials to build their games. I stopped to check in with Eddie and his partner. Eddie was tearing apart their creation and tears were welling up in his eyes. I instantly had an urge to step in and help him so I could see his smiling face again; however, reality set in and I reminded myself the importance of failure and building resilience. Rather than asking what was wrong and how could I make it better, I let him know it was OK to be frustrated, and I asked questions to help *him* find a solution: "What isn't working? Why do you think?"

I asked Eddie to think back to the activities done over the previous days. I asked him, "What contact points have you connected? What conductors and insulators are you using?" He pinpointed a problem with his design and immediately identified a change he could make to the design. This was such an important moment for Eddie as a learner. He continued working through the process, despite the struggles, and was eventually successful. See the photo on page 198.

**UDL Connection**

*Principle III: Engagement*
*Guideline 9: Self-Regulation*
*Checkpoint 9.2: Facilitate personal coping skills and strategies*

Once the games were complete, the students presented their creation and invited other groups to attempt their "operation" game and give feedback. As I circulated around the room, I heard students using their academic language as they tried their peers' games: "If you do that, you complete the circuit." "This part is an insulator." They were also using what they knew to ask questions and evaluate the designs: "Why isn't it working? Where is the contact point?" "Why is this part covered in foil? It's not making contact with the battery or bulb." This latter commentary revealed that while students demonstrated a better understanding of conductors and insulators, they didn't consider whether it was

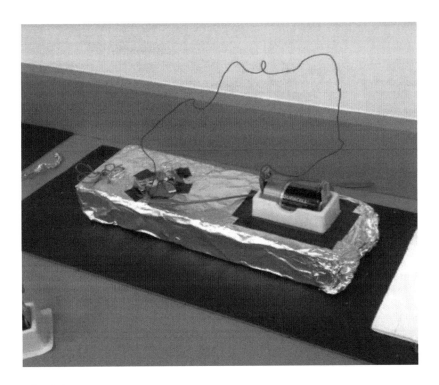

**Eddie's finished design.**

appropriate to use conductors for decorative parts of their designs. For example, like Eddie and his partner, some groups used foil to cover the base of their designs. This observation was important in help-ing students consider revisions to their designs.

As we ended the week and prepared to move on to a deeper examination of the flow of energy, we wrapped up by revisiting some of the ideas that had emerged during the students' discus-sions about the flow of energy in their circuits.

> **Teaching Tip:** Students are highly motivated to try out their peers' games. Another option would be to invite another class in to test the games and give their feedback.

As a class, we made a list of ideas and questions they had about energy. This would allow both the students to revisit and reflect on their thinking and me to revisit and reflect on my plans for the next lesson and to identify any adaptations that would be beneficial for students.

The students may have finished building their games, but they were not finished explaining how they worked. The comments that students had shared at this point revealed that many were confusing the transfer of *energy* and the flow of *current* (movement of the electrons) in the circuit—so I knew it would be important to distinguish between the two concepts in the next lesson. Their games would provide an ideal vehicle for examining energy transfers and exploring how they could use the energy in their battery to produce sound and motion, as well as light!

## Unpacking UDL: Barriers and Solutions

While reading through this vignette can help give you a sense of what happened during instruction, I would like to emphasize the planning that occurred before implementation. I had specific learners in mind as I planned the unit, and I identified the barriers that the activities I chose might pose for particular students, as well as solutions to address those barriers. I didn't need to come up with entirely new activities, but I was able to think about how I could enhance the activities to reduce the barriers and make the experience effective for all students.

Table 9.2 summarizes the connections between the Universal Design for Learning (UDL) framework and the overall design of this lesson. Following that, I describe a student in my classroom and how I considered his needs, interests, and abilities during the planning process.

| TABLE 9.2. UDL Connections | |
|---|---|
| **Connecting to the Principles of Universal Design for Learning** | |
| *Principle I. Representation* | |
| *Guideline 1: Perception* | |
| Checkpoint 1.3. Offer alternatives for visual information | In the Engage phase, the teacher reads the problem scenario aloud to support struggling readers. |
| *Principle II. Action and Expression* | |
| *Guideline 5: Expression and Communication* | |
| Checkpoint 5.1. Use multiple media for communication | For reluctant writers, provide a choice to use images or words to communicate ideas. Give an opportunity to use video (Screen Chomp) to share their completed circuits. |
| Checkpoint 5.2. Use multiple tools for construction and composition | Provide sentence starters, picture prompts, and word banks for students who need extra support to start writing. |

*(continued)*

| TABLE 9.2. UDL Connections *(continued)* | |
|---|---|
| **Guideline 6: Executive Functions** | |
| Checkpoint 6.2. Support planning and strategy development | Provide a step-by-step guide of the task to facilitate task completion for students who may lose focus or struggle to recall what to do. |
| Checkpoint 6.3. Facilitate managing information and resources | Discuss how to organize findings and/or provide a graphic organizer for data collection and organization. |
| | Provide tasks in a written format for students who struggle to recall information. |
| | Have graphic organizers to facilitate the design process to help students with focus and task completion. |
| **Principle III. Engagement** | |
| **Guideline 8. Sustaining Effort and Persistence** | |
| Checkpoint 8.3. Foster collaboration and community | Create strategic partner for peer interactions and support. |
| **Guideline 9. Self-Regulation** | |
| Checkpoint 9.2. Facilitate personal coping skills and strategies | Provide teacher feedback for managing frustration and persistence for task completion. |

### Learner Profile: Eddie

"Eddie" was a student who was capable in all academic areas, but collaboration was a barrier for him, as he often preferred to work alone and did not like to compromise. When activities were multistep or complex, Eddie could easily get frustrated as he had difficulty simultaneously recalling key concepts previously learned and integrating them in newer activities. When he did not get his way or he failed, he often shut down and did not work or participate.

I anticipated that Eddie's biggest struggle would be during the Evaluate phase. This phase required him to recall critical ideas that had been previously learned and apply them to a complex problem situation. Further, he was required to work with a partner. To help Eddie be successful, I was proactive in designing my lesson and included the following:

- A graphic organizer for Eddie to plan his game
- Thoughtful partnering to include him with peers likely to support his participation
- Frequent teacher check-ins and monitoring of discussions

Although these adaptations were specific to Eddie, I was also able to use the graphic organizers for other students. Eddie was not the only student in my class who benefited from more structured support to complete a task. By giving advance consideration to how I grouped students, I was able to identify particular supports and provide them.

---

### Questions to Consider

➢ To what extent did the activities in this lesson align with the purpose and intent of each phase of the 5E Learning Cycle? Could you envision other activities that would be appropriate for each phase?

➢ Were you able to follow the sequence of activities and the ideas that students developed across all five phases? What role did discussion play in the progression of the storyline?

➢ In what ways was the teacher able to assess students at the end of each day's session? How did this inform her instruction in the next phase of the lesson? What information did she elicit that could inform her instruction in the next learning cycle?

➢ As you read through the lesson, did you come across any activities that might pose a barrier for your own students? What principles of UDL might you apply in those instances?

---

## References

Keeley, P. 2008. *Science formative assessment: 75 practical strategies for linking assessment, instruction, and learning.* Thousand Oaks, CA: Corwin Press.

NGSS Lead States. 2013. *Next Generation Science Standards: For states, by states.* Washington, DC: National Academies Press.

# CHAPTER 10

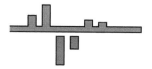

# Can You Hear Me *Know?*

### *Brooks Ragar*

M any students in my class play an instrument, so for our sound unit, we focus on trying to explain why different instruments sound different from one another—for example, a cello and a violin, a bass guitar and an acoustic guitar. It's a personally relevant phenomenon that allows students to share their experiences and culture through examples of instruments they play or ones that are featured in music that is important to them.

This chapter describes a lesson that provides a first introduction to sound and helps students distinguish between two related aspects of sound that are often confused—pitch and volume. Students observe how sound is caused by vibrations, and they explore how when more force is applied, the sound becomes louder. They also discover that when vibrations are faster, the sound becomes higher in pitch. A virtual oscilloscope is used to help students represent these sound properties of pitch and volume using a wave model. Figure 10.1 (p. 204) displays a conceptual storyline of this lesson.

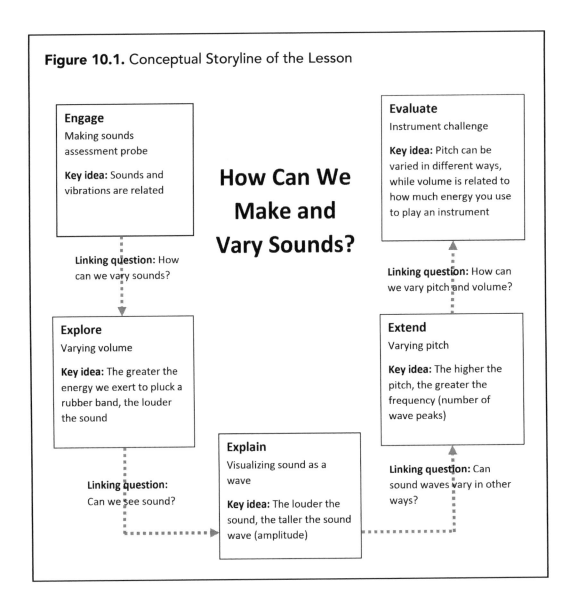

**Figure 10.1.** Conceptual Storyline of the Lesson

**Engage**
Making sounds assessment probe

**Key idea:** Sounds and vibrations are related

**Linking question:** How can we vary sounds?

**Explore**
Varying volume

**Key idea:** The greater the energy we exert to pluck a rubber band, the louder the sound

**Linking question:** Can we see sound?

# How Can We Make and Vary Sounds?

**Explain**
Visualizing sound as a wave

**Key idea:** The louder the sound, the taller the sound wave (amplitude)

**Evaluate**
Instrument challenge

**Key idea:** Pitch can be varied in different ways, while volume is related to how much energy you use to play an instrument

**Linking question:** How can we vary pitch and volume?

**Extend**
Varying pitch

**Key idea:** The higher the pitch, the greater the frequency (number of wave peaks)

**Linking question:** Can sound waves vary in other ways?

## LESSON VIGNETTE

I have not always enjoyed science. When I was in school, it was my least favorite subject. I was not a strong reader, and that was the only way my teachers "taught" science—by having students read about it in the textbook. I thought to myself, "When I become a teacher, I will not teach science this way!"

I had been teaching for 22 years, and I felt I was doing a much better job teaching science than how I was taught. I had moved from the textbook readings that characterized my own learning to hands-on activities, but little did I know just *how much better* my science teaching could be.

When I had the opportunity to attend the Quality Elementary Science Teaching (QuEST) program, I thought that spending two weeks in a campus dorm in July was not my idea of fun. However, two of my colleagues were going, so I thought, "Why not?" It turned out to be one of the best decisions I have ever made and the best professional development I have ever attended!

Experiencing learning science using the 5E Learning Cycle opened my eyes to a new and better way to teach science than what I was currently teaching—one that was not just hands-on, but also minds-on. I can't thank Dr. Hanuscin enough for inspiring me and teaching me how to become a better teacher. Thank you to Dr. van Garderen for introducing the principles of Universal Design for Learning (UDL) and explaining the importance of *always* considering the barriers that children might encounter in a lesson and how I can implement strategies so all students can have a successful learning experience. The opportunity to work with other QuEST attendees to implement these two approaches in a summer science camp for students really cemented things for me and prepared me to implement the 5E and UDL in my own classroom.

In this lesson, I focus on some basic ideas about sound that can be pretty tricky for students—pitch and volume. I used a virtual oscilloscope to help students visualize these two concepts using their own voices to produce images of sound waves. When I engaged my students with this technology, combined with hands-on explorations, they became just as hooked on science as I was in QuEST!

See Table 10.1 (p. 206) for alignment of the lesson to the *Next Generation Science Standards* (*NGSS*; NGSS Lead States 2013).

| TABLE 10.1. *NGSS* Alignment | |
|---|---|
| **Connecting to the *NGSS*—Standard 4-PS3: Energy** | |
| *www.nextgenscience.org/dci-arrangement/4-ps3-energy* | |
| • The chart below makes one set of connections between the instruction outlined in this chapter and the *NGSS*.<br>• The materials, lessons, and activities outlined are just one step toward reaching the performance expectation listed below. | |
| **Performance Expectation 4-PS3-2.** Make observations to provide evidence that energy can be transferred from place to place by sound, light, heat, and electric currents. | |
| **Dimensions** | **Classroom Connections** |
| *Science and Engineering Practices* | |
| *Planning and Carrying Out Investigations*<br><br>Make observations and measurements to produce data to serve as the basis for evidence for an explanation of a phenomenon. | Students investigate how sound can be produced and varied, and they observe these variations as waves using an oscilloscope. |
| *Disciplinary Core Ideas* | |
| *PS3.A: Definitions of Energy*<br><br>The faster a given object is moving, the more energy it possesses. (4-PS3-1)<br><br>Energy can be moved from place to place by moving objects or through sound, light, or electric currents. (4-PS3-2, 4-PS3-3) | Students observe objects vibrating (moving back and forth) and relate the height of the sound wave to the amount of energy used to produce the sound. |
| *Crosscutting Concepts* | |
| *Cause and Effect*<br><br>Cause and effect relationships are routinely identified, tested, and used to explain change. | Students make sounds in different ways and observe both the effect of their actions on the sound they hear and the appearance of the sound wave that represents the sound. |

**Teaching Tip:** The *NGSS* includes a performance expectation for first graders to plan and conduct investigations to provide evidence that vibrating materials make sound and sound can make materials vibrate. However, I do not want to assume that students were taught this or can recall it. Because it's a foundational idea for examining pitch and volume, I need to elicit their prior knowledge.

## Engage Phase

Many of my science lessons begin with one of Page Keely's formative assessment probes as a way to make assessment a seamless part of the lesson. These probes are designed to engage students in articulating their ideas as well as explaining their thinking, which we can revisit throughout the lesson to see if those ideas have changed. The particular probe that I chose for this lesson is "Making Sound" (Keeley, Eberle, and Farrin 2005). It provides a list of items that make sounds (e.g., a drum, rustling leaves, a popped balloon) and asks students to indicate which items involve vibrations. Space is then provided for the students to explain the "rule" they used to decide which items to check.

At the start of this lesson with my students, I asked them to get out their colored pencils or pens before I passed out the probe. We had established a routine of using one color to indicate their initial ideas and another color when they revisited the probe throughout the lesson. This allowed me (and them) to keep track of and visualize how their ideas had changed.

Once the students responded to the probe individually, I asked them to share and compare their ideas with their table group. To do this, I used a card sort (Keeley 2008), in which I provided each small group with index cards featuring the same objects listed on the probe and a T-chart on a large piece of chart paper. One column said "Involves vibrations" and the other said "Does not involve vibrations." The students used this to communicate their ideas with the rest of the class—noting where they had areas of

> **Teaching Tip:** I got in the habit of using colored pens and pencils when I noticed that students were erasing their initial ideas to write the correct ones. Being able to reflect on how their ideas changed is an important part of the sense-making process, so I want them to be able to recall what they thought beforehand. I communicate to students that I *expect* their ideas to change—after all, if they knew everything already, I wouldn't need to teach the lesson!

disagreement by taping the index card overlapping both sections of the T-chart. This was of particular benefit for my students who struggled with fine motor skills used in writing but also for those who had reading difficulties.

**UDL Connection**

*Principle I: Representation*
*Guideline 2: Language and Symbols*
*Checkpoint 2.5: Illustrate through multiple media*

**UDL Connection**
*Principle II: Action and Expression*
*Guideline 5: Expression and Communication*
*Checkpoint 5.1: Use multiple media for communication*

I enjoyed listening to the students' reasoning as they shared with one another, because this helped me evaluate what I was planning to do next and identify connections I could make to students' prior knowledge. When students realized that their peers' ideas were different from theirs, it provided the motivation to find out which ideas were correct! As they turned to me for an answer, I smiled and responded, "I guess we will have to investigate this tomorrow!" And with that I heard many dramatic sighs and the statement "You always say that, you never tell us the answers!"

As I reviewed students' responses, I noticed that several items they didn't recognize involved vibrations (e.g., rustling leaves, snapping fingers), and I made a note for us to revisit them later in the lesson so that the students could figure it out for themselves, rather than have me tell them the answer. Because all students recognized guitar strings as involving vibrations, I felt confident about going ahead with the exploration I had planned!

> **Teaching Tip:** This is a much more meaningful way to frame the activity as opposed to what I might have done previously—which was to provide them with some background information about sound. To me this is a better way for the students to take ownership of their learning by testing their own ideas against evidence.

## Explore Phase

As science time approached the next day, the students reminded me that I had said we were going to continue our exploration of sound. I loved to see the excitement and hear it in their voices! I explained to them that they would be building an instrument to help them explore sound. Before beginning our exploration, I shared what the specific question was for this particular activity that would help them answer our big question: "What do you think will happen to the sound that a rubber band makes when you change how hard you pluck the rubber band?" Knowing the routine, the students immediately got out their "predicting pens" to jot down what they thought they would find as they sought an answer to this question.

I wanted my students to have a shared experience that would enable them to make direct comparisons with one another, so I gave them the following instructions for making an instrument in the Explore phase:

1. Take a toothpick to poke a small hole in the bottom of a paper cup.
2. Use scissors to cut the rubber band. Tie one end of the rubber band to the toothpick.
3. Thread the rubber band through the hole in the paper cup so that the toothpick is inside the cup. The toothpick should hold the rubber band in place as you pull on the rubber band from the other side of the cup.
4. Place the cup upside down on a desk or table. Place the ruler so that it is standing upright against the cup. Use masking tape to tape the ruler to the cup.
5. Stretch the rubber band and tape it to the top of the ruler.
6. Keep the cup and ruler on the desk or table. Hold the cup in place with one hand and then pluck the rubber band.

I shared the materials we would be using, reminding students that they were for scientific purposes, not recess, and I reviewed the safety procedures (see the "Materials and Safety Notes" box).

---

### Materials and Safety Notes

**Materials**

| | |
|---|---|
| Goggles | Nonlatex rubber bands |
| Toothpicks | Plastic rulers |
| Paper cups | Masking tape |
| Scissors | |

**Safety Notes**

1. All involved must wear indirectly vented chemical splash goggles or safety glasses with side shields during all phases of these inquiry activities (setup, hands-on investigation, and take-down).
2. Direct supervision is required during all aspects of this activity to ensure that safety behaviors are followed and enforced.
3. Make sure that any items dropped on the floor or ground are picked up immediately after working with them—a slip/trip fall hazard.
4. Use caution when working with sharp objects (e.g., toothpicks, scissors). They can cut or puncture skin.
5. Follow the teacher's instructions for returning materials after completing the activity.
6. Wash your hands with soap and water after completing the activity.

---

**UDL Connection**

*Principle III: Engagement*
*Guideline 8: Sustaining Effort and Persistence*
*Checkpoint 8.3: Foster collaboration and community*

> **Teaching Tip:** Allowing students to make their own sound instrument is a time when the teacher's patience is crucial! I have each student make his or her own instrument; however, they are in table groups so they can help each other. I find I need to assist as well if I notice students becoming frustrated. Strict adherence to the directions is not important—having a working instrument is. I emphasize to students that the successful instrument is one that can make a sound. I provide the students time to explore making sounds with their instruments on their own before we move on together as a class.

Once my students had put together their instruments and had had time to explore making sounds with them on their own, it was time to collect data together in a more systematic way. I asked my teacher assistants to pass out the recording and data sheets that went along with the activity.

I told the students that we would vary the amount of force we used to pluck the rubber bands. They had already learned about the concept of force in a previous unit, so I asked them what they thought this might mean. They easily recognized that this meant the same as "how hard" they plucked their instruments. We decided to compare a gentle, medium, and strong pluck. I then asked them which would take more *energy* for them to do—a gentle pluck or a strong pluck. They recognized that the strong pluck involved more energy (which was enough for us to build on later without going into further depth at this time).

Because it could be hard to distinguish what the students were hearing if everyone was playing their instruments at the same time, we did this part together. Students described what they heard each time in small groups and then recorded their observations in a table in their notebooks (see Figure 10.2).

**Figure 10.2.** Sample Student Data Table From the Explore Phase

| Strength of Pluck | Observation (Sample Answer) |
|---|---|
| Normal pluck | "As I plucked the rubber band, it vibrated back and forth. It sounded like a guitar string." |
| Gentler pluck (less force than first pluck) | "I used less force to pluck the rubber band. It did not vibrate as much. The vibrating sound was not as loud as the first time I plucked the rubber band." |
| Harder pluck (more force than first pluck) | "I used more force this time. It vibrated a lot. The sound produced was louder than the first time I plucked the rubber band." |

At this point in the lesson, I was listening in on students' conversations and noting the kinds of observations they were making. Sometimes, the students needed prompting to notice specific things. For example, if a student indicated that his or her instrument "sounds like a guitar," I responded by saying, "Is it making sound in the same way? Tell me more about that." Or, I prompted them to look as well as listen. "What do you see when you pluck the rubber band?" I asked.

Some students may have used terms in their observations that we hadn't yet discussed. For example, if a student described the *volume* as being greater, I would ask that student to share his or her observation with the class, and I asked others to put in their own words what they thought that student meant by "volume." I find it is so important to really listen to your students! This is how I can gauge if they are on the right track or if they have some ideas that are inconsistent with their evidence.

As an example, one of my students explained *volume* as being higher or lower—which might sound like he was confusing volume and pitch, a common misconception. However, when I asked the student to explain what he meant, he described turning up the volume on his television (making it louder). Other students pointed out that their instruments didn't quite sound alike, and when pressed, they described this as "higher" or "lower" as well—but they made an analogy to playing different notes. I encouraged the students to record these observations so that they could make sense of them in the next session, and I made a mental note to revisit this example of pitch as we differentiated pitch and volume in the Extend phase of my lesson plan.

Before we wrapped up, I had the students revisit their predictions from earlier, and to make any changes. I asked them to respond to several questions as an exit slip for me to review before our next activity:

1. How did the sound your instrument made change as you changed the amount of energy you used to pluck the rubber band?
2. How does your observation compare with the results you expected?
3. Use your observations from the activity to form an explanation about how you think your device made sound.

As I reviewed the responses, I noted that almost everyone recognized that the harder they plucked their rubber band, the louder it became—but there were a few students who simply described the sound as getting higher or lower. From this, it was not clear to me whether they were referring to pitch or volume.

For many students, this experience was a confirmation of their expectations, rather than a surprise. Interestingly, each of the students had associated the sound of their instrument with vibrations—signaling that this particular activity allowed them to see and notice the vibrations as they were happening. A few students recognized that they transferred energy to the rubber band, but not many—this was something missing from their explanations that we would focus on integrating next!

## Explain Phase

"Yesterday you *heard* sounds," I began in this phase of the lesson. "But what if we could *see* sounds?" Students' attention was piqued by this question. For this part of the lesson, I used a virtual oscilloscope (see *https://academo.org/demos/virtual-oscilloscope*), a tool that provides a visual representation of a sound as a wave. I first demonstrated how the tool works by repeating the same observations that we did the day before—a gentle, normal, and strong pluck to one of the instruments that the students built—and asking students to watch how the wave changes. I instructed them to talk in pairs about what they had noticed, and then they reached a consensus as a group about what had occurred: "The harder I pluck, the louder the sound, and the taller the wave becomes."

> **Teaching Tip:** If we don't have a consensus, we will often retest so that students can verify what they are observing.

Now was the time to associate some scientific terms with what the students were describing in their own words—so I incorporated the science vocabulary into their thinking. I waited until this phase to incorporate the vocabulary because the students now had some concrete experiences and examples that they could connect with. "The height of a sound wave, or how tall it is, is called the *amplitude*," I explained. "When you *amplify* something, you make it louder."

"Like turning up the volume," a student offered.

"Yes—*volume* refers to how loud or soft a sound is," I confirmed.

"Is it like an amp, then?" one of my students interjected. "My brother plays electric guitar and when he plugs in his amp it gets really loud and my mom doesn't like it." The class erupted in giggles.

"Great connection!" I exclaimed. "How tall do you think the sound wave would be if he played his guitar for us?"

"Off the screen!" the student joked.

I pointed to the sound wave I had captured using the "freeze input" feature on the virtual oscilloscope, and I traced my hand along the peaks and valleys. "Notice this wave goes up and down," I began.

"Just like a vibration!" a few students called out.

"Did your rubber bands also go up and down?" I asked. Students referred back to their notebooks and pointed out that the rubber band "vibrated more" the harder they plucked it. When pressed about what "more" meant, the students weren't clear. Did it mean faster vibrations? Bigger vibrations? I sent the groups back to their tables to do a little more investigating.

As I walked around, students were pointing out that to pluck harder, they were pulling back or pushing down more on the rubber band—making it stretch and vibrate over a bigger space back and forth, different from when they do so gently. As we came back together, students agreed that when they plucked harder, the vibration got "taller," similar to the wave that represented the louder sound it made.

To help students make sense of this, I asked them what would take more energy—moving their arm back and forth a little bit, or moving it back and forth in a big motion? The students agreed that making a bigger motion required more energy. I then asked, "So, the more energy you used to pluck your rubber band, the larger a vibration it made, and …?" "The taller the wave!" students called out.

I then showed the students an image of two sound waves—which were the same in every way except for their height (see Figure 10.3)—and asked them what they could tell me about the two sounds just by looking at the waves. I asked the students to talk in pairs before sharing out. They offered responses such as the following:

- "One is louder than the other."
- "The tall one is louder."
- "It took more energy to make the tall one."

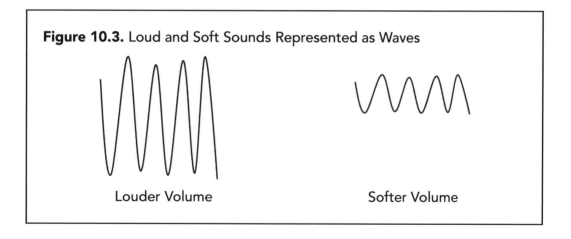

**Figure 10.3.** Loud and Soft Sounds Represented as Waves

Louder Volume          Softer Volume

I also pointed out that the two waves had the same number of peaks or humps—so that students noticed that the height was the only way they differed.

Because we had covered a lot of terminology here, I wanted students to be able to keep all of the terms straight. I used "foldables" as a way for the students to include these terms in their science notebooks. We made a "door" or "tab" style in which the word was written on the outside and it unfolded to reveal the definition. We included the following terms:

- *Volume*—how loud or soft a sound is
- *Amplitude*—how tall a sound wave is

**Teaching Tip:** I always try to have more than one way to present information to my students. This phase is also a great time to bring in nonfiction books or videos. I find that because they have experiences and vocabulary, students will be able to link these to what is mentioned in the books and videos.

## Extend Phase

Students, as I noted earlier, often confuse "pitch" (how high or low a sound is) with "volume" (how loud or soft a sound is). Because they were already familiar with waves that differed in height, in this phase of the lesson I extended their thinking to consider two waves that were the same height, but that differed in another way—the number of peaks (see Figure 10.4). I showed the students the two waves and asked them what they noticed and how they thought the two sounds might differ. This sparked a lively debate, because while the students suspected the sounds were different, they saw that the height was the same and therefore thought the two sounds were equally loud! I noticed some eager hands and exclamations of "I know! I know!" but realized that not everyone had had enough time to think about the question. I asked the students to think-pair-share: "What are other ways sounds can differ besides how loud they are?"

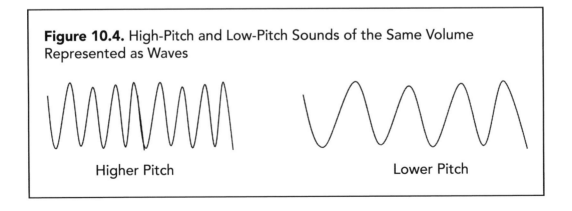

**Figure 10.4.** High-Pitch and Low-Pitch Sounds of the Same Volume Represented as Waves

Higher Pitch

Lower Pitch

When we came back together as a class, it was clear there were more students ready to share. Two hypotheses were put forward: (1) Something different was making the sound (similar to how two people have different voices), and (2) the sounds were higher or lower (similar to different notes on a musical scale). Several students pointed out that these were actually similar ideas because voices can also be high or low. I was delighted when my students made connections to what they were learning in music class about different instruments and choir roles (like alto or bass).

"How could we test those ideas?" I asked. Some students suggested using our voices to make high and low sounds, while others pointed to our class chime, which plays three different notes. Though this wasn't what I had initially planned, I recognized that it was a perfect tool to use! I displayed the virtual oscilloscope and readied the chime. "How should I strike it?" I asked. I was pleased when students pointed out that I should use the same amount of force. "Why is that important?" I probed. I pointed at the two waves we were examining, and that helped students identify that the two waves should match in height—so that meant the same loudness, which meant the same energy put into striking the chimes.

To students' delight, the waves displayed on the oscilloscope were similar to the example waves! "Which wave represented the *higher* sound?" I asked. "And which represented the *lower* sound?" I repeated the demonstration, giving the students another opportunity to distinguish this. As they recognized the difference, I explained that "high" and "low" in this case referred to the *pitch* of a sound—just as they commented before about people's voices being higher or lower pitch. I explained that we could imagine the peaks of the waves passing us, and the more peaks that pass, the more *frequently* the wave peaks pass. Scientists call this property of sound waves *frequency*.

I supported the students with a mnemonic device:

- "The higher the sound, the higher the number of peaks or frequency."
- "The lower the sound, the lower the number of peaks or frequency."

We then added these terms to our foldables as well:

- *Pitch*—how high or low a sound is
- *Frequency*—the number of peaks of a wave

## Evaluate Phase

For this final part of the lesson, I set up stations around the classroom with laptops that had the virtual oscilloscope ready for students to use. I pointed out that in the earlier part of the lesson (the Explore phase), one of the groups had noticed that their instruments were making sounds that were higher than another group's. I asked the students what they thought might have led to that difference. "We noticed that we had skinny rubber bands, and they had fat ones," they offered. Another group pointed out that they had heard different pitches, too, but it had happened when they stretched their rubber band tighter.

"It sounds like you are up for a challenge!" I announced, much to their delight. I provided the students with a handout of pictures of four different sound waves (see Figure 10.5). The students were tasked with describing the sound that each wave represented, relative to the others, in terms of pitch and volume. Then, they were challenged with producing the matching sound on their instrument and explaining how they accomplished that. Finally, the students verified their interpretation of the sound wave by using the oscilloscope while producing the sound.

As the students worked on the challenge, I found that they needed very little intervention from me—aside from a few well-posed questions when they were getting stuck. I asked, "What would happen if you stretch the rubber band tighter?" or "What would happen if you make your rubber band looser?" or "What if you used a thicker or thinner rubber band?" This sent them scurrying to adjust their instruments and try again. I checked in with the groups, asking them to demonstrate to me what they were discovering and recording on their tables, and listening to them use the vocabulary we had just reviewed.

Once all the groups had completed the activity, the students convened as a class to compare their findings. While all groups were able to produce the corresponding sounds (after some trial and error), what interested them the most is that they did so in different ways—there wasn't just one way to change the pitch of their instrument. This made me think they would enjoy our upcoming lessons in which we would explore different instruments and they would design their own!

As a final sense-making activity, we revisited students' initial ideas. I pointed out to them that while their rubber bands vibrated when they made sounds, there were some sounds that they thought did not involve vibrations. "What do you think we would see if we made those sounds with our virtual oscilloscope on?" I asked. Their excitement was obvious, as they figured out what they would be doing next. I had made a point to gather materials ahead of time that they might need to test this—such as leaves to rub together to simulate rustling leaves, rocks to tap against each other, and a balloon to pop. Other examples, such as snapping fingers, are more easily tested! I asked each group to test out any of the examples that they thought were not caused by vibrations. I then asked them to revisit their ideas and to indicate any that they had changed their minds about.

**Figure 10.5.** Student Sound Challenges

| Sound Wave | Describe the sound it shows | How did you make the sound with your instrument? | What did the wave look like? |
|---|---|---|---|
|  | Pitch:<br><br>Volume: | | |
| 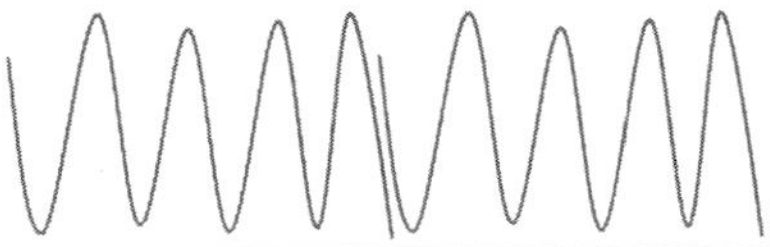 | Pitch:<br><br>Volume: | | |
|  | Pitch:<br><br>Volume: | | |
|  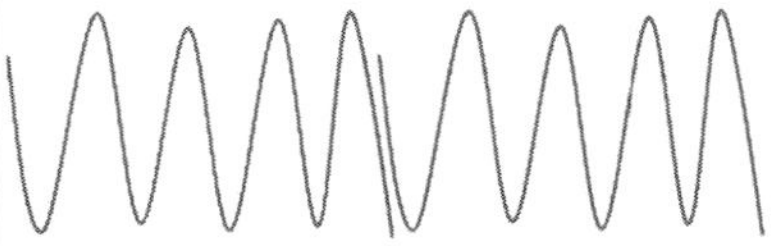 | Pitch:<br><br><br>Volume: | | |

Although the students seemed to have been successful in differentiating pitch and volume, I knew that they would need further encounters with those concepts, just as they would need to revisit their ideas about the relationship between vibrations and sound. As we moved on in our unit, we would get closer to the *NGSS* performance expectations. We would explore different ways to produce and vary sound, and we would also build on the ideas about energy that students had just begun to develop in relation to sound. Our science journey was far from over!

## Unpacking UDL: Barriers and Solutions

Table 10.2 (p. 218) summarizes the UDL principles, guidelines, and checkpoints that I applied when designing the activities for each phase of the 5E Learning Cycle lesson to meet the *general* needs of the students in my classroom. Following that are some examples of how I identified barriers and strategized solutions to meet the *specific* needs of three of the learners in my classroom, specifically those who pushed in or were pulled out for special education instruction.

| TABLE 10.2. UDL Connections | |
|---|---|
| **Connecting to the Principles of Universal Design for Learning** | |
| **Principle I. Representation** | |
| **Guideline 2: Language and Symbols** | |
| Checkpoint 2.5. Illustrate through multiple media | In the Engage phase of the lesson, I use a card sort of pictures of objects that were listed on the original probe in writing, to ensure that all my students, particularly the struggling readers, know what the objects are. |
| **Principle II. Action and Expression** | |
| **Guideline 5: Expression and Communication** | |
| Checkpoint 5.1. Use multiple media for communication | Rather than writing answers on a T-chart, something a number of my students struggle to do, I create T-charts on large pieces of paper for each group during the Explore phase. Using the cards from the card sort, the students then "record" their answers on the T-chart. Whenever writing is expected, my students Shawna, Derrick, and Erin are given an option to record their responses using drawings and labels. |
| Checkpoint 5.2. Use multiple tools for construction and composition | Erin experiences difficulty writing, and to prevent her from falling behind and as a way to check her understanding, I scribe for her. |
| **Principle III. Engagement** | |
| **Guideline 8: Sustaining Effort and Persistence** | |
| Checkpoint 8.2. Vary demands and resources to optimize challenge | Some of my students need the demands of a task reduced in order to be able to participate. For Shawna and Derrick, to prevent frustration, I premade their instruments. |
| Checkpoint 8.3. Foster collaboration and community | I create table groups with students of varying strengths and abilities so that if one or more students experience difficulty in a task (such as reading), another student could read for them. |

**Learner Profile: Derrick and Shawna**

I had two students, "Derrick" and "Shawna," who came into my classroom just for science. Both these students had been identified as having an intellectual disability. When they came to my room, they did not have an aide. Both students were three or more years behind in reading and struggled to comprehend what was being taught. However, they were very excited about science and enjoyed coming to class each day.

The strengths these two brought to my classroom were the willingness to participate in class discussions, an eagerness to learn, and the ability to work well with others! Throughout this lesson, I made changes for Derrick and Shawna to be successful in their learning. But some of the solutions that I used were unique to their needs. These included the following:

- In the Explore phase, I made their sound instruments. This helped with frustration following multiple directions and steps to make them. And I always made extra just in case more students had difficulty completing the task.
- In the Evaluate phase, these students had the option of illustrating and labeling their response as opposed to a written response.

**UDL Connection**

*Principle III: Engagement*
*Guideline 8: Sustaining Effort and Persistence*
*Checkpoint 8.2: Vary demands and resources to optimize challenge*

**UDL Connection**

*Principle II: Action and Expression*
*Guideline 5: Expression and Communication*
*Checkpoint 5.1: Use multiple media for communication*

There was one procedure that I used with Derrick and Shawna in mind but that I also used with other students as a general solution for those who struggled with reading and writing (such as Erin, in the next learner profile). It was as follows: I sat them together with a table group of two other students who were higher functioning. In the Engage phase, rather than completing the task as a written response, I changed this probe into a card sort that the table groups could work on together. Further, the students worked together to complete a T-chart with a chart for the card sort.

### *Learner Profile: Erin*

"Erin" was a student of mine who had a learning disability. She went in and out of the classroom the entire day for different types of support—particularly related to different areas of literacy. I worked with the special education teacher to make sure that she was not pulled out during science. Thankfully, the special education teacher understood the importance of science and was always willing to work with me. Erin's reading level was far below anyone in my class, and she struggled with writing complete sentences. She loved the hands-on experiments—something I strived to include in my learning cycles as a motivator for her (and others!) to engage in the content—but the challenge was in checking for understanding.

To accommodate Erin's specific needs, when and where necessary, I used the following solutions:

- Erin would draw what she knew, and as a way to check for understanding, she then would tell me about it.
- To check for understanding and because writing was difficult, at times I scribed for her.

**UDL Connection**

*Principle II: Action and Expression*
*Guideline 5: Expression and Communication*
*Checkpoint 5.1: Use multiple media for communication*

**UDL Connection**

*Principle II: Action and Expression*
*Guideline 5: Expression and Communication*
*Checkpoint 5.2: Use multiple tools for construction and composition*

## Questions to Consider

➢ To what extent did the activities in this lesson align with the purpose and intent of each phase of the 5E Learning Cycle? Could you envision other activities that would be appropriate for each phase?

➢ Were you able to follow the sequence of activities and the ideas that students developed in the lesson? How did the storyline of the lesson progress? How did the teacher cope with concepts that students might potentially confuse?

➢ In what ways was the teacher able to assess students during each phase of the lesson? How did this inform future instruction?

➢ In what ways did the solutions that the teacher identified meet the needs of the specific students spotlighted in this vignette? In what ways did they benefit all students? Could you think of other solutions you might use for your own learners?

## References

Keeley, P. 2008. *Science formative assessment: 75 practical strategies for linking assessment, instruction, and learning.* Thousand Oaks, CA: Corwin Press.

Keeley, P., F. Eberle, and L. Farrin. 2005. *Uncovering student ideas in science: 25 formative assessment probes.* Arlington, VA: NSTA Press.

NGSS Lead States. 2013. *Next Generation Science Standards: For states, by states.* Washington, DC: National Academies Press.

# CHAPTER 11

# Seeing the Light

### *Deborah Hanuscin*

Our ability to see is important to our everyday lives; however, it's not something we typically approach with a scientific mindset to understand how vision works. Yet, studying how we see is inherently relevant to students' everyday lives, making it an ideal phenomenon to investigate! The particular lesson in this chapter has been used with inservice and preservice elementary teachers, and it is a redesign of a lesson originally taught to elementary school students. The conceptual storyline of this lesson (see Figure 11.1) is outlined on page 224.

**Figure 11.1.** Conceptual Storyline of the Lesson

## How Do We See?

**Engage**
Formative assessment probe

**Key idea:** We have different ideas about how we see

**Linking question:** Which ideas are correct?

**Explore**
Light-box activity

**Key idea:** We can see inside the box some times and not other times

**Linking question:** Why could we see in some situations and not others?

**Explain**
Data analysis of conditions in which we can see or not see

**Key idea:** We can only see when light is present

**Extend**
Reflecting-light activity

**Key idea:** Light must be able to reflect off an object and enter our eyes for us to see it

**Linking question:** Is light the only thing we need to see?

**Evaluate**
Self-assessment and model

**Key idea:** Scientists' ideas may change with new evidence

**Linking question:** How have our ideas changed?

## LESSON VIGNETTE

My elementary school teaching preceded the release of the *Next Generation Science Standards (NGSS)*, so I have worked hard to improve my ability to design and enact instruction that reflects the *NGSS* (NGSS Lead States 2013). Some of that has involved developing *NGSS*-aligned learning experiences for teachers in professional development programs, for elementary students, and for preservice teachers (Hanuscin and Zangori 2016; Hanuscin, Arnone, and Bautista 2016).

This is all part of my commitment to "practicing what I preach" as a science educator. Being able to do so not only provides opportunities for me to model dimensions of practice to teachers, but also helps them *experience* what it is like to learn science when taught in this manner. The teachers who have attended our Quality Elementary Science Teaching workshops over the years have reiterated what a powerful experience it is—and the sense of accomplishment that comes from *learning with understanding* as opposed to being told information or attempting to memorize it. It helps the teachers realize that everyone is capable of learning when barriers are removed and support is provided.

Although I have implemented this lesson in several different contexts and with different audiences, I have written it from the perspective of teaching it with elementary students, re-creating actual interactions I've had with them. Thus, it represents an amalgam of experiences and highlights what I felt to be particularly challenging or interesting encounters throughout.

Please see Table 11.1 (p. 226) for alignment of the lesson to the *NGSS*.

| TABLE 11.1. *NGSS* Alignment | |
|---|---|
| **Connecting to the *NGSS*—Standard 4-PS4: Waves and Their Applications in Technologies for Information Transfer** | |
| *www.nextgenscience.org/dci-arrangement/4-ps4-waves-and-their-applications-technologies-information-transfer* | |
| • The chart below makes one set of connections between the instruction outlined in this chapter and the *NGSS*.<br>• The materials, lessons, and activities outlined are just one step toward reaching the performance expectation listed below. | |
| **Performance Expectation 4-PS4-2.** Develop a model to describe that light reflecting from objects and entering the eye allows objects to be seen. | |
| **Dimensions** | **Classroom Connections** |
| *Science and Engineering Practices* | |
| *Developing and Using Models*<br>Develop a model to describe phenomena. | Students develop a model to explain how light enables them to see an object. |
| *Disciplinary Core Ideas* | |
| *PS4B: Electromagnetic Radiation*<br>An object can be seen when light reflected from its surface enters the eyes. | Students examine conditions under which objects can and cannot be seen, relating this to the amount of light present and the ability for light to enter their eyes. |
| *Crosscutting Concepts* | |
| *Cause and Effect*<br>Cause and effect relationships are routinely identified. | Students analyze patterns in their data to identify what conditions enable them to see or not see an object. |

## Engage Phase

To start the lesson and elicit students' ideas about how we see, I used a version of the formative assessment probe "Apple in the Dark" (Keeley 2019). I prepared a response handout for students, but I set the stage for the task by using an associated video that I found online in which a teacher provides the scenario of students venturing deep into a cave, turning out their lantern, then trying to see the apple they brought with them (see *www.youtube.com/watch?v=wtuPE8TryrQ*). Because I knew that many of my students enjoyed camping and exploring our local caves, I knew this would capture their attention! Just as I did at the beginning of each lesson, I made sure that my microphone was working and asked for a thumbs-up from students to indicate that they were able to hear me as well as the audio from the speakers.

Students watched the video with captions on as the narrator sets the stage and poses a question about whether they would be able to see an apple in a dark cave in which no light entered. "You might already have some ideas, so let's see which idea you agree with most," I said. Then I projected the handout they had been given using our document camera and read aloud so that all my struggling readers could access the information:

1. You will not be able to see the apple at all.
2. You will be able to see the apple after your eyes adjust to the dark.
3. You will see only a faint outline of the apple after a while.
4. You will see the apple after a while, but it will not look red.

Beneath these options were instructions for students to circle the idea they agreed with most, then explain their reasoning.

**UDL Connection**

*Principle I: Representation*
*Guideline 1: Perception*
*Checkpoint 1.2: Offer alternatives for auditory information*

**UDL Connection**

*Principle I: Representation*
*Guideline 1: Perception*
*Checkpoint 1.3: Offer alternatives for visual information*

As students worked silently (with a few reminders to do so) on writing their responses, I circulated around the room noting which ideas they had circled. I checked in with my student Ainsley, who had begun writing her response with the knowledge that her special education teacher would help her rewrite it for correct spelling and syntax during our language arts time, so it was OK if it was not perfect now. I asked her to tell me about her thinking, and I was confident that she was prepared to share her ideas orally with the group.

**UDL Connection**

*Principle II: Action and Expression*
*Guideline 5: Expression and Communication*
*Checkpoint 5.3: Build fluencies with graduated*
*levels of support for practice and performance*

I then checked in with each table captain (a role that rotated daily to provide opportunities for all students to learn how to work with others) and asked that when their table was finished, they (or a member they choose) place a sticky note next to the options each member of their group had selected before they went on to discuss what they thought. This would create a "sticky bars" graph (Keeley 2008) on our whiteboard at the front of the room through which students could compare their ideas. My student Alan was team captain today, and he was just a few feet from the board, but I was confident he could judge for himself whether he felt able to post his group's selections. I continued to drop in on the various groups, listening to their thinking and noting whether and how they recognized the importance of light to their ability to see.

**UDL Connection**
*Principle III: Engagement*
*Guideline 8: Sustaining Effort and Persistence*
*Checkpoint 8.3: Foster collaboration and community*

After the groups had ample time to share, I called the class back together to examine our graph. Interestingly, we had a number of students who had chosen each of the responses! "It looks like we have some disagreement about which idea is correct," I began. "Shall we settle it by a vote? By rock-paper-scissors?" My kidding was met with a chorus of *Nos* and a few good-natured eye rolls as well. My students were well aware at this point in the year that not only did I have a corny sense of humor, but also that scientific claims had to be tested against evidence. We had discussed that unlike elections, scientific disagreements aren't settled by a vote. Arm wrestling, on the other hand, piqued their interest, and I noted my student Nick and his elbow partner starting to engage in that. "Nick, how did you know I was joking just now?" I asked and smiled, regaining his attention. "Scientists do experiments, they don't arm wrestle," he offered. "Well, maybe they do arm wrestle, but not when they are doing science," he added with a grin.

I turned back to the rest of the class and said, "OK, then, what should we do in our next science lesson?" The resounding chorus of "Test it!" conveyed their enthusiasm for what would come next. I told the students to hang onto their papers, and I took a photo of our sticky bars graph as we wrapped up the session.

## Explore Phase

The next day when science time rolled around, students were eager to begin. "Are we going to go in a cave?" asked Alan, clearly concerned about how he would get there in his wheelchair.

"Actually," I explained, "I have brought the cave to *you!*"

I had created several light-tight cardboard boxes (Ashbrook 2012) that resembled caves, and I showed these to the students. I provided one to each group with the instructions to inspect them and let me know whether they thought these would work as a stand-in for a cave and why. My motive was both to help students identify important features of our caves (an opening at one end for them to look into and no other way for light to enter) and to help them get over the novelty of this new object and their (understandable) desire to observe and touch it. See the "Materials and Safety Notes" box.

**UDL Connection**

*Principle II: Action and Expression*
*Guideline 4: Physical Action*
*Checkpoint 4.1: Vary the methods for response and navigation*

---

## Materials and Safety Notes

### Materials

Light-tight cardboard boxes

Bags of plastic fruit:

- Red apple
- Green apple
- Yellow apple
- Banana

Clear plastic window

Wax paper window

Thick fabric window

Flashlights

Small mirrors

Small targets

### Safety Notes

1. All involved must wear indirectly vented chemical splash goggles or safety glasses with side shields during all phases of these inquiry activities (setup, hands-on investigation, and take-down).
2. Direct supervision is required during all aspects of this activity to ensure that safety behaviors are followed and enforced.
3. Make sure that any items dropped on the floor or ground are picked up immediately after working with them—a slip/trip fall hazard.
4. Make sure that all fragile items are removed from the activity zone before the lights are turned off.
5. Remember to walk carefully during the activity when the lights are off to help prevent trip-and-fall accidents.
6. Use caution handling mirrors if they are glass or metal. They can be sharp and can cut or puncture skin.
7. Never eat any food or food facsimile used in science activities.
8. Follow the teacher's instructions for returning materials after completing the activity.
9. Wash your hands with soap and water after completing the activity.

I provided each group with a bag of plastic fruit: a red apple, a green apple, a yellow apple, and a banana. I then demonstrated the next steps of our exploration with one group serving as my assistants. Taking turns, each person would secretly choose *one* of the objects to place inside the cave (box), then pass it around for everyone to look into to determine whether they could see the object. We acted this out with the students trying their best not to reveal to the rest of the group what they saw (or didn't see) when they looked inside.

I then showed the students a feature that many of them missed—a flap at one end of the box that could be opened—and I showed them several "windows" that could be inserted in place of the flap. One was made of clear plastic, one of wax paper, and another of thick fabric. I explained to the class that they would repeat this process in three more rounds, using each of the covers. I then provided each group with a data sheet (see Figure 11.2) and instructed the team captains to make sure they had a recorder and that everyone took turns and shared their ideas.

**Figure 11.2.** Group Data Sheet for the Explore Phase

| Test | What we could each see: |
|---|---|
| Flap closed | |
| Clear plastic | |
| Wax paper | |
| Thick fabric | |

**UDL Connection**

*Principle II: Action and Expression*
*Guideline 6: Executive Functions*
*Checkpoint 6.3: Facilitate managing information and resources*

**UDL Connection**

*Principle III: Engagement*
*Guideline 8: Sustaining Effort and Persistence*
*Checkpoint 8.3: Foster collaboration and community*

**Teaching Tip:** Team captains are the only people allowed to come to me or get my attention when their group needs help—this cuts down on the number of people who may be up or roaming around during the activity, for safety's sake, and it helps me more easily keep track of which groups I need to check in with.

Most groups were able to complete their observations with the box before it was time to clean up. I offered two groups a chance at recess or during morning work the next day to finish up the last round they had left. One of the groups suggested that they could work on it during lunch time, which I indicated was fine—as long as they didn't mistake the plastic fruits for their lunch! This elicited a grin from Nick, who promised not to eat our science project.

## Explain Phase

We started the next day with each group taking out their data tables. I had written four questions on the whiteboard for students to discuss as a group before they compared their findings:

1. Under what conditions could you see the object?
2. Under what conditions could you not see it?
3. Could you tell the color of the objects you saw?
4. What does your group think determines whether you can see an object?

I offered my student Lara the option for her group to move to the back table where it was a bit quieter so they could all hear each other, something I knew was difficult for Lara when there was a lot of background noise. I walked around listening carefully to the students' conversations and looking over their data charts. Where I noticed some anomalies (students indicating they could see the object and identify it when the flap was closed), I asked them to show me how they tested it. This helped them identify an error in their recording, which they were able to correct. At another table, some students saw a red apple but one reported she saw a green apple—my immediate reaction was to assume that this was a mistake, but I realized there was a chance that students were just guessing when they were unable to tell.

**UDL Connection**
*Principle III: Engagement*
*Guideline 7: Recruiting Interest*
*Checkpoint 7.3: Minimize threats and distractions*

I brought the students back together after their small-group discussions to share what they had found. As the first group began, I prompted other groups by asking, "Does that agree with what your group thinks? How does their idea compare to yours?" For this particular activity, students' findings seemed to corroborate with one another, so we quickly reached a consensus—something that wasn't always the case!

I then challenged the groups to come up with a "one-sentence summary" (Keeley 2008) that communicated what they had learned in their investigations. I provided sentence strips to each group so that all students could focus on the content as opposed to the process of writing, which some struggled with, and we posted them on the board to compare. The sentences included the following:

- "We can't see in the dark."
- "We can see if there is light."
- "The more light, the better we see."
- "Light is needed for us to see."
- "You can see in the light, not the dark."
- "It's harder to see when it's darker."

**UDL Connection**

*Principle II: Action and Expression*
*Guideline 5: Expression and Communication*
*Checkpoint 5.3: Build fluencies with graduated*
*levels of support for practice and performance*

The sentences that each group came up with revealed that there was not as much consensus as I had initially thought; several seemed to leave open the possibility of seeing without there being a source of light. I made a note that the ability to identify sources of light was an important element in the conceptual storyline that I may have overlooked! The students, however, did not note this subtle difference in their statements, and they agreed. I pushed them to think a bit further by asking the last group, "Would it ever get dark enough that you couldn't see?" Unexpectedly, this prompted students to raise the point that sometimes you can't see when there is too much light—like the sun in your eyes. I was unsure about how to move the discussion forward, so I decided to let the students sit with these ideas a bit longer. I wondered if that was the right decision.

We wrapped up this part of the lesson with a drawing activity. I asked the students to think about when they were able to see the piece of fruit *and* could tell its color, and then to draw a picture of how they think that had happened. "How can you explain, in pictures, how you were able to see the fruit?" I inquired. As I heard a few students express concern about their drawing abilities, I added that they could caption their picture and use words to help explain what they had drawn. I collected the students' drawings to review before the next session, and I noticed that Alan's page had more words than pictures while Ainsley's had mostly pictures.

*Principle III: Engagement*
*Guideline 7: Recruiting Interest*
*Checkpoint 7.1: Optimize individual choice and autonomy*

## Extend Phase

While many students grasped that we need light to see, I was not convinced that that reflected a deeper understanding that the light reflects off an object and enters our eyes. A review of their drawings and captions revealed that some students thought that light shines *from* our eyes to see, while others had used arrows that either showed the direction they were looking or perhaps that light was coming from their eyes. Still others recognized that light was shining *onto* objects, but not that it was reflecting off of those objects.

For this activity, I provided pairs of students with a flashlight and a small mirror. Around the room, I hung small targets (the familiar red and white circles) near each group's table. I asked for a student volunteer to be my partner so I could demonstrate the instructions, and I picked Nick because I knew this would help him understand the directions too. One partner would stand in such a way that he or she could not see the target (such as with one's back to it). They would then hold the small mirror, while their partner had the flashlight. The goal was to work together so that the first person could see the target using the mirror—the challenge was that the classroom lights would be turned off!

"I wonder where I should shine the light," I said aloud. "On my partner? At the mirror? On the target?" A chorus of student whisperings revealed that they had some ideas already. I emphasized that the students should work in their area near their target (and not wander around the darkened classroom) and that they should take turns in each role with their table partner.

**UDL Connection**

*Principle II: Action and Expression*
*Guideline 5: Expression and Communication*
*Checkpoint 5.3: Build fluencies with graduated levels of support for practice and performance*

**UDL Connection**

*Principle III: Engagement*
*Guideline 8: Sustaining Effort and Persistence*
*Checkpoint 8.3: Foster collaboration and community*

As the students were working, I (carefully, using my own flashlight) moved about the room to check in on them. They were obviously enjoying the hands-on activity, but I wanted to ensure that it was minds-on as well, so I asked questions like the following:

- "What do you notice when you have the flashlight off? On?"
- "What happens when you shine the flashlight at the target? The mirror? Your partner?"
- "Where is the flashlight aimed when you can see the target best?"
- "Where do you think the light is going?"

After all the pairs had switched roles, I turned the lights back on and called the students back together—making sure that the materials were placed back in a bin at their tables to avoid distractions during our discussion. I asked the pairs at each table to compare notes, then work together to draw on their whiteboards how they were able to see the target.

As the groups begin drawing, I walked around and observed. For some groups, I asked them to label the parts of the drawing that were not clear, or to tell me about their ideas. My point was not to shape their thinking, but to help them communicate it more clearly. When all groups were finished, we held a "gallery walk" to compare ideas. All students pushed in their chairs, which left an outer aisle around all tables wide enough to allow the students to walk (or wheel, in Alan's case) around and view each other's work. We had done this before, so it was a familiar routine for the students, but nonetheless we reviewed our norms for quiet voices and moving patiently so that all could see. I circulated along with the students, so I had a sense when we were reaching the end of the walk. I could see students making their way back to their own desks, and I reminded them to do so quietly.

**UDL Connection**

*Principle II: Action and Expression*
*Guideline 5: Expression and Communication*
*Checkpoint 5.3: Build fluencies with graduated levels of support for practice and performance*

**UDL Connection**

*Principle III: Engagement*
*Guideline 7: Recruiting Interest*
*Checkpoint 7.1: Optimize individual choice and autonomy*

I had a poster in the room that featured a list of discussion starters that we used following gallery walks, such as the following:

- "I noticed that …"
- "I didn't understand …"
- "Why did you choose …?"
- "Ours was similar/different because …"

We used these to talk about what we saw in our gallery walk and compare the ideas of the class. One of the first things that students noticed was that everyone had used arrows in their drawings, but the arrows were facing different ways. Through a discussion, they realized that there were different meanings intended—some were showing the path the light was traveling (from the flashlight to the target, then to the mirror, and reflecting off into the person's eye), while others were showing the "line of sight" of the person—the direction he or she was looking.

What the students seemed to agree on was that they could see the target best when the flashlight was shone on the target. The fact that students already knew that light reflects off or bounces off a mirror allowed many to make the connection that the light was bouncing off the mirror and entering the eye; however, some were skeptical about the idea because when their partner shined the flashlight on the mirror and it reflected into their eyes, they were unable to see. I did my best to ask them whether they were trying to see the target or the flashlight, but this idea was not fully resolved. I then pulled down one of the targets and asked the students how we might check whether the light was reflecting off of it. This produced some confused looks, so I rephrased it: "If the target can reflect light, then what do you think will happen if I shine the light on this target while I am holding it near the whiteboard?"

I asked the students to think-pair-share. Afterward, a few volunteers suggested that the light would be reflected onto the whiteboard. I asked a helper to dim the lights and then we tried it out. I could see a faint red on the whiteboard, but I realized that the students couldn't see this—so I had them repeat this at their tables, using the fruit from our earlier investigation and a flashlight. Using the different colors allowed them to notice some differences—and to understand that the colors were coming from the light reflecting off the fruits.

I noticed that students were starting to put together their ideas, but they still needed some time to process. We set the whiteboards aside to pick up our discussion in the next session. The students needed to extend their thinking further!

The next session I began by asking the students if they had any more ideas or questions about the models that we had sketched out the previous session. What a difference a little thinking time makes! The students raised some very important questions. The first one was "Can the light reflect more than once?" To help answer this, I gave students a challenge—the reverse of what we did yesterday! Each group would try to hit their target with their flashlight, but after bouncing off of *two* mirrors. The students were visibly triumphant as their groups found that they could do this. They agreed that, based on this evidence, light *can* reflect more than once.

Another question that the students had was "Does it only work with a flashlight?" I thought that this question was important, because it signified that the students were considering the source of light. I asked, "Why do you think we turned off the lights when we used the flashlight?" An audible chorus of *Oh!*s let me know that the students had made a connection.

I closed by asking the students to individually redraw their own ideas about how we see objects. Before they drew, they clarified that they would use arrows to signify the direction that light was traveling (as opposed to showing the direction in which the person was looking).

## Evaluate Phase

As a final sense-making activity, I asked the students to revisit their original responses to the assessment probe from the beginning. We repeated the procedure by viewing the video clip again and reviewing the answers (particularly to support learners such as Nick and others who needed support in both reviewing and identifying key ideas), and then each student took a colored pen from the supply caddy to change or add to what they had written before. Again, I asked the team captains to pick a member to place a sticky note next to the response that each of their group members had chosen.

> **UDL Connection**
> *Principle I: Representation*
> *Guideline 3: Comprehension*
> *Checkpoint 3.4: Maximize*
> *transfer and generalization*

> **Teaching Tip:** Having students use a different colored pen at the end of the lesson helps make clear whether and how their ideas have changed.

I then shared the photo of our original sticky-bars graph and projected it next to the new one on the whiteboard. "Some of you have changed your ideas," I pointed out. "Do you think scientists ever change what they think? Talk about that with your group." I provided the students with this small-group discussion to ensure that everyone had enough time to process their ideas before we had a whole-class discussion. I was pleased when students pointed out that the evidence didn't match their ideas and that scientists might experience this too. A few even pointed out that people once thought the Sun went around the Earth, but now we know the Earth goes around the Sun.

Just as scientists aren't easily convinced to change their ideas, I had a few students who still didn't seem to be fully convinced of what they had seen (or not seen) with their own eyes. I knew that it would be important to revisit this idea and connect it to other ideas they were learning. I decided to follow up this lesson and begin the next with the formative assessment probe "Can It Reflect Light?" (Keeley 2012), as a way to identify the next steps in my instruction. I was hoping that the students would realize that if they could see something, it was reflecting light, but I expected that several may select items such as water, a mirror, glass, or shiny metal from the list and not check clouds, wood, or soil. I was confident, though, that they would be just as eager to test those ideas as they were in this lesson, and that we would continue to develop the model they had constructed so far. I knew that this one lesson wasn't enough for them to demonstrate understanding of the *NGSS* performance expectation, but it had gotten them closer!

## Unpacking UDL: Barriers and Solutions

Table 11.2 summarizes the Universal Design for Learning (UDL) principles, guidelines, and checkpoints that I had applied when designing the activities for each phase of the 5E Learning Cycle lesson to meet the *general* needs of learners in the classroom. Following that are some examples of how I identified barriers and strategized solutions to meet the *specific* needs of several hypothetical students, based on actual students with whom I had worked.

| TABLE 11.2. UDL Connections | |
|---|---|
| **Connecting to the Principles of Universal Design for Learning** | |
| *Principle I. Representation* | |
| *Guideline 1: Perception* | |
| Checkpoint 1.2. Offer alternatives for auditory information | In the Engage phase, captions were made available on the video to ensure that all learners, particularly Lara who had a hearing impairment, had access to the content. |
| Checkpoint 1.3. Offer alternatives for visual information | Printed information, such as the statements in the Engage phase, were displayed and read aloud to ensure that all students could access the information. |

*(continued)*

| TABLE 11.2. UDL Connections *(continued)* | |
|---|---|
| **Guideline 3: Comprehension** | |
| Checkpoint 3.4. Maximize transfer and generalization | In the Evaluate phase, students were given the opportunity to revisit and review information to strengthen the links between key ideas that they had learned. |
| **Principle II. Action and Expression** | |
| **Guideline 4: Physical Action** | |
| Checkpoint 4.1. Vary the methods for response and navigation | To provide equal interaction in the cave activity during the Explore phase, particularly for Alan who was in a wheelchair, an alternative cave was made. |
| **Guideline 5: Expression and Communication** | |
| Checkpoint 5.3. Build fluencies with graduated levels of support for practice and performance | Ainsley struggled with writing. To support her, the special education teacher provided additional time and instruction with writing. But during class time, Ainsley was given an option to share her thinking orally, as provided in the Explain and Extend phases. |
| | During the Explain phase, all students were provided with prewritten sentences to discuss and compare as opposed to expecting them to write statements, which can be challenging for some learners. |
| | Prior to starting an activity in the Extend phase, a demonstration of how to do the activity was provided to ensure that all students, particularly Nick who had difficulty following directions, understood the task to be able to fully participate. |
| **Guideline 6: Executive Functions** | |
| Checkpoint 6.3. Facilitate managing information and resources | A group data sheet was provided during the Explore phase to help learners who struggle with organizing information and to ensure that everyone has the data required for the next phase of the activity. |

*(continued)*

| TABLE 11.2. UDL Connections *(continued)* |
|---|
| **Principle III. Engagement** |

| **Guideline 7: Recruiting Interest** | |
|---|---|
| Checkpoint 7.1. Optimize individual choice and autonomy | During the Explain and Extend phases, students were given a choice for the response format (drawing or writing), allowing for all students to demonstrate what they have learned and promote connection to their learning. |
| Checkpoint 7.3. Minimize threats and distractions | Background noise could be distracting for Lara, making it difficult to follow conversations. Having the option to move to a quieter area during group discussions, such as the group discussion during the Explain phase, provided a way to vary the level of background noise. |
| **Guideline 8. Sustaining Effort and Persistence** | |
| Checkpoint 8.2. Vary demands and resources to optimize challenge | Nick's learning expectations were modified (or differentiated) to meet his needs as a way keep him motivated and engaged for successful task completion. |
| Checkpoint 8.3. Foster collaboration and community | For groupwork activities during the Engage, Explore, and Evaluate phases, students are provided with roles (e.g., team captain) or responsibilities (e.g., hold the mirror) that are rotated to ensure that all students get an opportunity to develop social interaction skills as well as to build a collaborative environment for learning. |
| | During the Explore phase, students work together to complete the activity and one student is assigned to keep the data for the group. |

### Learner Profile: Alan

"Alan" had muscular dystrophy, and the medication he took had caused him to gain weight, which had exacerbated issues he was having with physical mobility. He used a wheelchair when he needed to travel more than a few yards. He was self-conscious about the weight gain and the misperception that his weakness was simply due to being too large to move quickly. Nonetheless, he enjoyed school and exceled in all areas academically, particularly demonstrating a talent and passion for writing.

To support Alan in his learning, I made sure that he would not be excluded from an activity, creating an alternative "cave" as opposed to visiting a real cave during the Explore phase of the lesson.

### Learner Profile: Ainsley

"Ainsley" was a relatively quiet student who enjoyed playing sports. She often blended into the background during class, and she rarely spoke up in whole-class discussions. She was self-conscious about her academic performance, and the struggles she encountered with writing in particular, especially when it came to spelling. Ainsley was still using "invented" spelling for many words. This was related to a learning disability, for which she had an individual education plan (IEP) and received assistance from a special education teacher during language arts instruction. Despite her difficulties with writing, she was able to express herself clearly and effectively orally.

During the lesson, there were several times that students were required to write. Knowing that this was particularly difficult for Ainsley, I included opportunities for her to orally explain her thinking to me and/or to record her ideas using pictures and captions. She also worked with her special education teacher to expand and improve on what she had written during her science lessons. At times, such as during the Explain phase, I wrote out information so that this did not present as a barrier for completing the task.

### Learner Profile: Nick

"Nick" was short in relation to his peers, but full of more energy than most. He had a happy disposition but got easily frustrated with complex tasks. He had been identified as having a mild intellectual disability (with an IQ around 70), and his IEP included support and modifications in all academic areas. Though he received instruction in the general education classroom, he received support, along with Ainsley, from a special education teacher who pushed in.

Nick was self-conscious, and he had expressed on more than one occasion that he was "not smart." He would often resort to goofing off as a way to get attention, rather than academics, but he had a keen sense of humor. While Nick's learning expectations were modified in that he was not expected to learn the material to the same degree as his classmates, I still fully expected him to participate in the learning cycle and to develop understanding about the same content as his peers.

**UDL Connection**
*Principle III: Engagement*
*Guideline 8: Sustaining Effort and Persistence*
*Checkpoint 8.2: Vary demands and resources to optimize challenge*

Supports for Nick included ensuring group participation, such as being given a role or responsibility for group work activities that occurred throughout the lesson; an option to use pictures rather than writing his ideas; a review of materials and key ideas such as the opportunity to rewatch the video during the Evaluate phase; and a demonstration of how to complete an activity. These supports were provided throughout the learning cycle.

### Learner Profile: Lara

"Lara" had a hearing impairment that affected her ability to understand what was being said without amplification. She wore a hearing aid, but she found it difficult to distinguish the speaker's voice when there was an abundance of background noise. This made receiving oral instructions and participating in discussions difficult, though she had no problems with her speech or oral expression. At all times Lara wore an amplification system in which the teacher wore a microphone that connected to her hearing aid, so she could hear what was being said. However, other solutions needed to be built into any lesson that I taught, to ensure that Lara had access to the content when it was not just the teacher speaking.

In this lesson, captions were provided along with the video so that Lara could read what was being said, and when working in groups, her group moved to a quieter section of the room to reduce or eliminate background noise.

---

## Questions to Consider

➤ To what extent did the activities in this lesson align with the purpose and intent of each phase of the 5E Learning Cycle? Could you envision other activities that would be appropriate for each phase?

➤ Were you able to follow the sequence of activities and the ideas that students developed in the lesson? How did the storyline of the lesson progress for the students? Where were areas of divergence from what the teacher had in mind? What key ideas were most difficult for students?

➤ In what ways was the teacher able to assess students during each phase of the lesson? How did this inform future instruction?

➤ In what ways did the solutions that the teacher identified meet the needs of the specific students spotlighted in this vignette? In what ways did they benefit all students? Could you think of other solutions that you might use for your own learners?

---

## References

Ashbrook, P. 2012. Shining light on misconceptions. *Science and Children* 50 (2): 30–31.

Hanuscin, D., and L. Zangori. 2016. Developing practical knowledge of the *Next Generation Science Standards* in elementary science teacher education. *Journal of Science Teacher Education* 27 (8): 799–818.

Hanuscin, D., K. A. Arnone, and N. Bautista. 2016. Bridging the "next generation" gap: Teacher educators implementing the *NGSS. Innovations in Teaching Science Teachers* 1 (1). *http://innovations.theaste.org/bridging-the-next-generation-gap-teacher-educators-enacting-the-ngss.*

Keeley, P. 2008. *Science formative assessment: 75 practical strategies for linking assessment, instruction, and learning.* Thousand Oaks, CA: Corwin Press.

Keeley, P. 2012. Seeing the light. *Science and Children* 49 (6): 28–31.

Keeley, P. 2019. Formative assessment probes: Apple in the dark. *Science and Children* 56 (5): 15–17.

NGSS Lead States. 2013. *Next Generation Science Standards: For states, by states.* Washington, DC: National Academies Press.

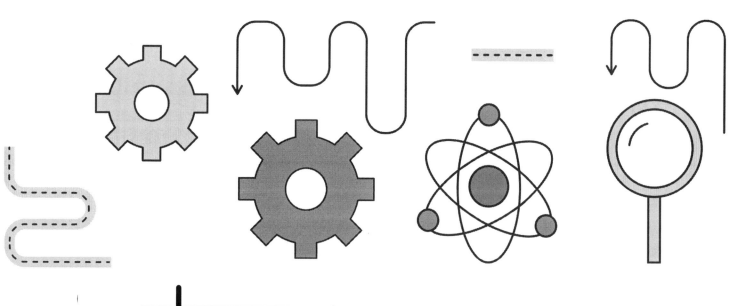
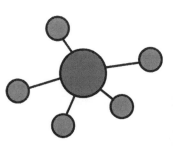

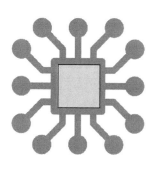

# PART III

## Applying the Frameworks

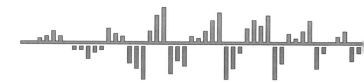

# CHAPTER 12

# Reframing Your Instruction

**Delinda van Garderen, Deborah Hanuscin,**
**Cathy Newman Thomas, and Kate M. Sadler**

Throughout this book so far, we have given you numerous examples of how the 5E Learning Cycle and Universal Design for Learning (UDL) work together as a way to plan for and teach science to your learners, but you may still be wondering where to start. There are no strict rules or formulas for implementation. The best advice we can give you is to pick an area of the framework that motivates you and to just start. However, this can be easier said than done. To help you identify a productive starting point, spend some time answering the following questions (p. 246) and thinking about what your next steps might be.

## STOP AND CONSIDER ...

What are some of the most memorable or intriguing ideas you've read about so far?

_____

_____

_____

_____

_____

_____

To what extent are you already applying these ideas in your own practice?

_____

_____

_____

_____

_____

_____

As you read, what aspects of your own science instruction did you find yourself thinking about in a critical way?

_____

_____

_____

_____

_____

Are there opportunities to incorporate some of the ideas into your science instruction in small steps? How?

_____

_____

_____

_____

_____

_____

So, now what? The rest of this chapter is divided into two main sections. The first aligns with Chapter 1 and focuses on the 5E Learning Cycle (including conceptual storylines and seamless assessment), while the second section focuses on UDL (Chapter 2). Both sections could be used together to develop an overall plan of your next steps, or you could choose to focus on one section and start there. For example, if you feel your instruction already closely aligns with the 5E, then you might want to begin with UDL and be working to support a particular learner in your classroom.

Within each section you will be provided an overarching framework of core ideas for implementation along with different ideas to help you get started. Again, we are not expecting you to start over from scratch, but to use these frameworks to enhance your instruction using your existing materials and resources. After all, you wouldn't go out and buy a new car when you have a flat tire, an empty tank, or a dead battery! Making even small adjustments to your existing lesson plans can make a big difference for your students' learning.

## A Note on Lesson Plans

"Lesson plans" come in many different forms and varieties. Your district may even have a format that you are required to use. It may seem inconsequential what format you use as long as you know what you are doing; however, we would suggest otherwise. The format you use can have a tremendous impact on the quality of the learning and your ability to evaluate student understanding. Importantly, it reflects an inquiry-based approach to instruction that is important not only for students who typically do well in science, but also especially for diverse learners (Scruggs, Brigham, and Mastropieri 2013). If you have a required lesson plan format, we encourage you to consider its alignment with the frameworks presented in this book:

- Does the format allow you to identify and include activities appropriate to each of the five phases of the 5E Learning Cycle?
- Does the format include structures to help support conceptual coherence?
- Does the format make explicit the opportunities for assessment that occur throughout?
- Does the format include a way to embed "solutions" for specific students *as you plan*?

If you answered "No" to any of these questions, we encourage you to identify any flexibility you may have in the format you are required to use and/or to consider how you might work within those constraints to apply these frameworks. In our experience, teachers have been able to advocate for changes to

incorporate these frameworks, in particular because of the research evidence that supports them.

## Reframing Instruction With the 5E Learning Cycle, Conceptual Storylines, and Seamless Assessment

The teachers we've worked with in the Quality Elementary Science Teaching (QuEST) program often comment that once they've learned about conceptual storylines, they can't help but view lessons through this lens. Accordingly, we think a great place to start is viewing your own instruction through the lenses of the 5E Learning Cycle, conceptual storylines, and seamless assessment. In Table 12.1, we outline a series of overarching questions that can serve as a guide in developing or reorganizing your planning to reflect the learning cycle, ensure a coherent conceptual storyline, and embed seamless assessment for learning throughout.

## Considering the Purpose of Activities in Each Phase of the 5E

You likely have a host of activities you've collected that relate to the concept you want to teach. But where should you put each in the sequence of the learning cycle phases? There is not one right answer! The same activity could be used in different phases, as long as it is used in a manner that reflects the *purpose* of the phase. Consider, for example, a card-sort activity that requires students to organize cards according to some particular prompt. This activity could be used for many different purposes based on the phase it is being used in. For example,

- *Engage*—as a way to determine what students understand about a concept or misconceptions they may have (e.g., put in order)
- *Explore*—as a way to group items (e.g., which items belong together)
- *Evaluate*—as a way to demonstrate what they have learned (e.g., sort which items belong to a particular term)

Throughout the vignettes presented in Part II of this book, you may have noticed that similar learning activities appear—but they don't always appear in the same phases. For example, videos were used by Christine Meredith (Chapter 4), Betsy O'Day (Chapter 8), and Deborah Hanuscin (Chapter 11). Revisit these chapters, if needed, to examine where a video appeared in the lesson sequence and what the purpose of using the video was.

## Beyond Teaching Topics

Think back to one of the vignettes you read. What was the lesson about? Chances are, you could use one word such as "magnets" or "matter." However, think

| TABLE 12.1. Three Lenses for Analyzing Your Lessons | | |
|---|---|---|
| **Alignment With the 5E Learning Cycle** | **Coherence of the Conceptual Storyline** | **Embedding Seamless Assessment** |
| *Engage* <br><br>• Does the activity focus on students' prior knowledge vs. introduce new information? <br>• Is there a central question or challenge to motivate students? <br><br>*Explore* <br><br>• Does the activity allow for student-directed/firsthand exploration of concepts or phenomena? <br><br>*Explain* <br><br>• Are the students responsible for explaining their ideas? <br>• Does the teacher help link new vocabulary and explanations to the activity and students' ideas? <br><br>*Extend* <br><br>• Does the activity require students to apply their ideas in a new context? <br><br>*Evaluate* <br><br>• Does the summative assessment task allow teachers to assess change in students' prior knowledge elicited in the Engage phase? | • Is there a central focus on *one* main concept or learning goal? <br>• Does the lesson make explicit the link between the hands-on activities and concepts or ideas? <br>• Does the idea in each activity build toward the idea in the next activity? <br>• How does the storyline of the lesson build on or link to lessons that come before and after? | • Does assessment occur before, during, and after instruction? <br>• Are the assessments embedded in the learning activities in meaningful ways? <br>• Are a variety of assessment strategies used? (both formal and informal) <br>• Are students self-assessing in addition to being assessed by the teacher? <br>• Does the assessment facilitate teachers' evaluation of students and evaluation of their teaching, and does it help identify next steps in instruction? |

about what the students were understanding or figuring out. Chances are, this can't be stated in one word but needs a complete sentence or question—for example, "Magnets can be used to solve problems" or "How can we solve problems using magnets?" As you look at your own lessons, it can be easy to assume you have coherence if you focus on topics alone (all the activities are about matter), when in reality you might be addressing multiple concepts or big ideas. The mantra "less is more" applies here. As you look at your own lesson, do you focus on one main concept or idea, or are you trying to address every concept related

to a topic? If your lesson is organized around a particular phenomenon, as is becoming more common since publication of the *Next Generation Science Standards* (*NGSS*; NGSS Lead States 2013), ask yourself what the underlying concept is that will help students make sense of the phenomenon. Similarly, if your lesson has an "essential question" as its organizer, is the concept implicit in the question? A conceptually coherent lesson builds on one main idea in depth. If you can't identify that, it's probably the place to start!

We like to think of the main concept as the "final chapter" of the storyline—the ending understanding that students will reach. The first chapter is … ? Wherever your students are starting! What is their prior knowledge? What are the likely ideas or misconceptions they hold that will be the stepping-stones to understanding the big idea? Building the storyline is about imagining how your students will move from where they are to where they will be when they reach the learning goals of the lesson. What ideas will they need to develop along the way? What experiences will help them do that? How will they get from one experience to the next? For inspiration, revisit the conceptual storyline maps that are included with each lesson vignette in Part II.

## It's a Learning *Cycle*, After All

Assessment for learning, or embedding formative assessment throughout the learning cycle, can help you identify students' readiness for next steps and make adjustments to your instruction to better support their learning. However, it's not just about the *next steps*—it's also about coming "full circle" and going back to the beginning. Try this: Look back at several of the vignettes and compare the Engage and Evaluate phase activities. To what extent do they connect? Are students circling back to revisit their initial ideas and reflect on how those have changed? Very often we get caught up in moving on to the next thing and we don't stop to consider *how far we've come*. Involving students in sense-making activities that help them become aware of their changing ideas is powerful for learning. This kind of self-assessment can help students feel more capable and more confident as learners!

## Identifying Your Next Steps

Ideally, we would encourage you to plan your science lessons using the 5E Learning Cycle; however, we recognize that this may be difficult if you have never planned this way before or implemented a 5E lesson yourself. To help you, we offer some smaller steps and goals that you might consider working toward:

- Narrow the focus of your lesson on one main idea or concept.

- Plan to embed formative assessments into your lesson.
- Invert the sequence of activities in your lesson to reflect the principle of "exploration before explanation" from the learning cycle.
- Generate a list of ideas that you liked from the vignettes and chapters that you could integrate into your existing curricula.
- Use or adapt one of the learning cycles presented in this book and implement it with your students.

Like us, you may find that enlisting the help of a colleague is a great way to support your mutual learning and to share your successes.

## Reframing Your Instruction With UDL: Planning to Include All Learners

In addition to understanding the key principles of UDL, it is important to understand the UDL can be applied to *all* aspects of our teaching and learning—the "curriculum." Broadly, all aspects of the curriculum, sometimes referred to as the four pillars of curricular components, include (1) goals, (2) methods, (3) materials, and (4) assessment (CAST 2018). Any one of these areas can pose a barrier to science learning for our diverse learners. Figure 12.1 provides details about each area.

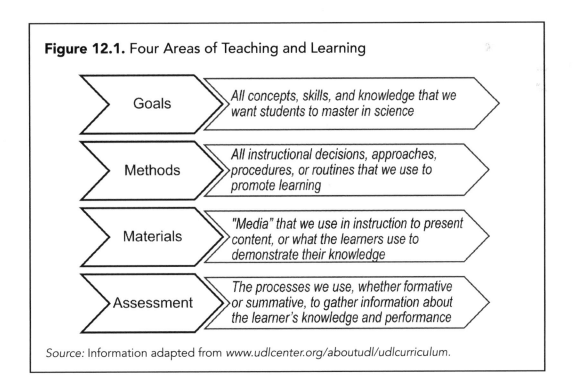

**Figure 12.1.** Four Areas of Teaching and Learning

| | |
|---|---|
| Goals | All concepts, skills, and knowledge that we want students to master in science |
| Methods | All instructional decisions, approaches, procedures, or routines that we use to promote learning |
| Materials | "Media" that we use in instruction to present content, or what the learners use to demonstrate their knowledge |
| Assessment | The processes we use, whether formative or summative, to gather information about the learner's knowledge and performance |

*Source:* Information adapted from *www.udlcenter.org/aboutudl/udlcurriculum.*

Ideally, the best time to apply the UDL framework is while you are designing, developing, and planning your science curriculum and lessons. The goal is to be as proactive as possible to prevent any missed learning opportunities due to inadvertent or unintended barriers to learning. We recognize that you may be using an existing curriculum that has lessons already planned for you. Often school districts and schools purchase and develop such materials to support teachers in meeting local, state, and national standards. However, even these materials need to be evaluated to identify potential barriers that your students may encounter so that you can, again proactively, identify solutions to embed in your instruction.

In Figure 12.2, we outline a series of overarching questions (or steps) that you can use to guide you as you implement UDL in your instruction, whether planning a lesson from the beginning or using materials already provided.

**Figure 12.2.** Overarching Questions for Planning With UDL

### UNIVERSAL DESIGN FOR LEARNING

**Representation:**
What barriers in the curriculum may prevent students from accessing the content they are to learn?

**Action and Expression:**
What barriers in the curriculum may prevent students from participating in the lesson and demonstrating what they know?

**Engagement:**
What barriers in the curriculum may prevent students from motivating, engaging, and sustaining interest in learning science?

√ What solutions (guideline and checkpoint) can you embed in your 5E Learning Cycle to help address these barriers?

√ Are the solutions varied and flexible to provide multiple ways to support student learning?

√ Do these solutions connect to student-identified strengths and challenges?

√ Do the solutions promote learning to support students to meet the learning goal?

## Considerations for Implementation

You may have noticed in the various chapters that we have made reference to the idea of planning for the *general* or *specific* needs of our learners. A starting point for implementing UDL in your instruction is anticipating and planning for general barriers that students in your class at-large or as a whole may experience. Examples: The content may be too easy or too difficult for the age group you are working with, there may not be enough space in the classroom to carry out a certain activity, there may not be enough opportunities for the learners to share their own ideas, or the assessment may be situated in a context that the students in your classroom cannot identify with.

Essentially, what is guiding this stage of implementation of UDL solutions is the need to embed instructional practices that reflect sound findings from research about learning and teaching science (National Research Council 2005). As a result, practices that would enhance opportunities and provide equitable access to learning for all the students in your room are embedded during planning.

### STOP AND CONSIDER ...

Throughout the vignettes in this book, we embedded numerous solutions that were designed to promote access to learning for all students. Find a vignette, and read through it. As you read it, highlight or underline solutions that were proactively anticipating general learning needs.

*Vignette read:* _____

What are some general solutions that you currently use or have learned from the vignettes that make sense to embed in your instruction? How might these solutions benefit your learners?

_____

_____

_____

_____

_____

_____

Another way to apply UDL solutions is to meet the need of a "specific" learner (e.g., a student with a learning disability, a student with a visual impairment, a student for whom English is a second language) or group of specific learners, such as students who struggle to write and students from culturally and linguistically diverse backgrounds.

As you plan solutions for specific learner needs, there are three important considerations to keep in mind. First, it is important to recognize that when you are thinking about specific learner needs, the barrier needs to be accurate. If the barrier is not correctly identified, the solution that you apply may not work. For example, you may observe that one of your students, "Charlie," is "unmotivated to learn" and then identify the barrier as the learner lacking motivation. Based on that barrier, you identify the solution for Charlie is "providing the choice of a reward for task completion."

**UDL Connection**

*Principle III: Engagement*
*Guideline 7: Recruiting Interest*
*Checkpoint 7.1: Optimize individual choice and autonomy*

However, after embedding this solution in a learning cycle, you found that it didn't work! Charlie still did not complete the task and continued to remain unmotivated and unengaged. What happened? Yes, this learner was at times "unmotivated," but that only occurred when he was expected to read text during the learning cycle. The reality is that Charlie struggles to comprehend what he is reading and "switches off" during reading activities, thus appearing unmotivated during any activity that requires independent reading. A more appropriate solution may have been connected to providing him with ways to comprehend the material.

**UDL Connection**

*Principle I: Representation*
*Guideline 2: Language and Symbols*
*Checkpoint 2.3: Support decoding of text, mathematical notation, and symbols*

Second, be careful assuming that for learners with the same or similar "labels" (e.g., students with learning disabilities or English language learners [ELLs]), their learning needs or characteristics will require, or respond in the same way to, a solution that you use. Further, it may be necessary to test several solutions that "fail" with a learner before you fully understand what the exact nature of the barrier is, or how your demand is affecting learning for an individual student or group of students. As you implement solutions, keep a record of what works and what does not work for each learner.

Third, when thinking of solutions, build on and connect to the strengths that the student has. It is easy to think that because a student doesn't have certain skills in place (e.g., the ability to decode words or comprehend math facts), this is where the focus should be. However, the same student may be extremely interested in and know a lot about a certain topic, which could be used in a proactive way that may powerfully motivate the student, provide other ways that the student could contribute to a group activity, or be used to further develop skills.

---

### STOP AND CONSIDER ...

Throughout the vignettes in this book, we embedded numerous solutions that were designed to promote access to learning for specific learners. Find a vignette, and read through it. As you read it, highlight or underline solutions that were proactively anticipating specific learning needs.

*Vignette read:* _____

What are some specific solutions that you currently use or have learned from the chapters that make sense to embed in your instruction? How might these solutions benefit particular students?

_____

_____

_____

_____

_____

_____

---

Importantly, solutions should reflect evidence-based practices—those that have been identified via systematic research to work with all learners or for specific types of learners. There are many websites and resources available that summarize evidence-based practices to use. We highlight some of them in Table 12.2 (p. 256).

**TABLE 12.2. Select Free Resources for Locating UDL Information and Solutions**

| Free Resource | Web Address | Brief Description |
|---|---|---|
| **CAST** | *www.cast.org* | CAST originated the UDL framework. This is the main website link. |
| UDL Guidelines | *http://udlguidelines.cast.org* | CAST provides guidelines along with explanations and additional links for solutions. |
| Publications and Media | *www.cast.org/whats-new/publications-media.html* | CAST provides access to books, journal articles, policy statements, podcasts, videos, and more. |
| Free learning tools | *www.cast.org/whats-new/learning-tools.html* | CAST offers free learning tools. |
| **CAST Professional Learning** | *http://castprofessionallearning.org* | CAST Professional Learning offers opportunities for educators to enhance their professional understanding of UDL. |
| Free UDL Resources and Tips | *http://castprofessionallearning.org/free-udl-resources-and-tips* | Here you will find quick tips, articles, and papers to help plan for and implement UDL in the classroom. |
| Free UDL Webinar Series | *http://castprofessionallearning.org/free-udl-webinars* | Webinars are offered on a variety of topics including learner variability, lesson design, and UDL implementation. |
| **The IRIS Center at Vanderbilt University** | *https://iris.peabody.vanderbilt.edu/module/udl* | The IRIS Center at Vanderbilt University is a free online resource for general educators who teach diverse learners. Resources include online modules, case studies, information briefs, and more on a variety of topics. |
| Online module | | Module: "Universal Design for Learning: Creating a Learning Environment That Challenges and Engages All Students." |

*(continued)*

| TABLE 12.2. Select Free Resources for Locating UDL Information and Solutions *(continued)* | | |
|---|---|---|
| **Free Resource** | **Web Address** | **Brief Description** |
| **The UDL Implementation and Research Network (UDL-IRN)** | *http://udl-irn.org* | The UDL-IRN is focused on increasing implementation and research on UDL. |
| Network and learn | *https://udl-irn.org/network-and-learn* | Free online webinars and discussion sessions are offered on a variety of UDL topics. |
| Special interest groups (SIGs) | *https://udl-irn.org/udlhe-network* | Existing SIGs include implementation and professional development. Interested people can form a new SIG with support from the UDL-IRN. |
| Resources | *https://udl-irn.org/home/udl-resources* | A series of documents are shared to support UDL implementation. |
| Newsletter | *https://udl-irn.org/weekly-updates* | A multimedia newsletter is published and archived. |
| **UDL supporting learners in British Columbia schools** | *www.udlresource.ca* | British Columbia schools in Canada offer materials, including a self-paced online course on UDL. |
| Self-directed course | *http://udlresource.ca/2017/11/a-self-directed-course* | This online course offers six individual modules, including one on implementation and another designed to extend teacher knowledge. |
| **National Science Teaching Association (NSTA)** | *www.nsta.org* | NSTA offers some materials on UDL. |
| Universal Design in Science Learning | *www.nsta.org/publications/news/story.aspx?id=51695* | This NSTA digest article shares ideas to integrate UDL into inquiry, including lab and field work. |

Finally, and while this may seem obvious, don't forget to ask the experts available to you in your school building or district. In many cases, these experts (e.g., the ELL teacher, the science coordinator or curriculum specialist, the special education teacher) may know many different solutions for you to use. Furthermore, they may already be successfully using solutions (such as supports for a bilingual student, access to newly released science resources, or

accommodations from an individual education plan) that you could also implement in your classroom.

## Making a "Cheat Sheet"

To keep track of and reduce the need to constantly find solutions, you might create a list or a solutions "cheat sheet" that you could readily turn to as needed. This sheet can be referred to "in the moment" of teaching if you need to make adjustments during teaching or as you plan your instruction. There are numerous ways you could organize your cheat sheet—for the class as a whole, for each group of learners, for each student, and so on. Table 12.3 provides a sample cheat sheet that could be consulted when planning a series of lessons.

| TABLE 12.3. Sample Solutions "Cheat Sheet" for Common and Frequent Barriers to Learning | | |
| --- | --- | --- |
| **Barrier** | **Potential UDL Solutions** | **Example Resources** |
| The learner has weak reading decoding and fluency skills.<br><br>• He or she reads so slowly, and with so many errors, that comprehension is poor.<br><br>This barrier is related to student characteristics. | Partner reading (Checkpoint 8.3) | Partner reading is a structure for using peer mentors and models to support decoding, fluency, and therefore comprehension (*www.readingrockets.org/ strategies/paired_reading*) |
| | Text to speech (Checkpoint 1.3) | NaturalReaders is a free text-to-speech option (*www.naturalreaders.com*)<br><br>Voice Dream Reader is a pay text-to-speech option, and it has a companion writing support (*www.voicedream.com*)<br><br>Learning Ally (*http://learningally.org*) and Bookshare (*www.bookshare.org/cms*) are free options for students who receive special education services of Section 504 support for a print disability |
| | Models (Checkpoint 3.3) | Paul Andersen video (*www.bozemanscience.com/ngs-developing-using-models*)<br><br>NSTA (*http://ngss.nsta.org/Practices.aspx?id=2*) |

*(continued)*

**TABLE 12.3. Sample Solutions "Cheat Sheet" for Common and Frequent Barriers to Learning (continued)**

| Barrier | Potential UDL Solutions | Example Resources |
|---|---|---|
| The key science concept is abstract.<br><br>• Science is characterized as being conceptually dense, and important ideas are often abstract.<br>• For example, from Chapter 2, while students can see the results of a proper circuit when energy is conducted, the term *conductivity* is abstract.<br><br>This barrier is related to science content. | Computer simulation (Checkpoint 2.5) | PhET Interactive Simulations (*https://phet.colorado.edu/en/simulation/conductivity*) |
| | Student demonstration (Checkpoint 5.2) | NSTA lesson on circuits (*http://ngss.nsta.org/Resource.aspx?ResourceID=509*) |
| | Examples and non-examples (Checkpoint 3.2) | |
| The learner's primary language is not English.<br><br>• Science literacy is a goal of the *NGSS*, so science learning materials include opportunities to listen, speak, read, and write about technical science vocabulary and concepts.<br><br>This barrier is related to learner characteristics and to science content. | Electronic translation tools (Checkpoint 2.4) | Google Translate (*https://translate.google.com*)<br><br>Prizmo (*https://creaceed.com/prizmo*)<br><br>iTranslate (*https://itunes.apple.com/us/app/itranslate-translator-dictionary/id288113403?mt=8*)<br><br>iTranslate Voice (*http://itranslatevoice.com*)<br><br>CAST Science Writer (*http://sciencewriter.cast.org/welcome*) |
| | Culturally relevant and responsive (Checkpoint 7.2) | *NGSS* and English language learners (*http://ell.stanford.edu/sites/default/files/pdf/academic-papers/03-Quinn%20Lee%20Valdes%20Language%20and%20Opportunities%20in%20Science%20FINAL.pdf*)<br><br>NSTA and Culturally Relevant Science Teaching Collection (*https://learningcenter.nsta.org/mylibrary/collection.aspx?id=2chjOK8Rsi8_E*) |
| | Embed support for vocabulary and symbols within text (Checkpoint 2.1) | Picture-Perfect Science Lessons (*www.nsta.org/store/product_detail.aspx?id=10.2505/9781935155164*) |

## Identifying Your Next Steps

Ideally, planning with UDL would be a regular part of your practice; however, we recognize that this may be difficult if you have never planned this way before and have not developed a repertoire of solutions that work for your students. To help you, we offer some smaller steps and goals that you might consider in working toward making planning with UDL a habit:

- Examine the learning goals for your learning cycle, and make sure they are accessible for all students.
- Develop a cheat sheet for students in your class.
- Take one lesson and build in solutions connected to one principle from the UDL framework.
- Build in one solution to a lesson for one, two, or three students in your classroom.
- Rewrite an assessment task to build in solutions for all the students in your classroom.
- Create a document of potential solutions to use when planning for some or all of your students.
- Examine a learning cycle or unit of instruction, and embed solutions for common barriers that a number of your students may experience.

## Closing Thoughts

We know that change can be hard, and it's OK to start small rather than over-haul. The teachers featured in this book and those we worked with in QuEST received a year of support from our staff, with continued support from their grade-level colleagues, as they worked toward improving their science teaching. We hope that the strategies we've provided in this chapter will help you begin to make small changes to your practice initially—and we believe you'll find they make a big difference in your teaching and in student learning.

## References

Center for Applied Special Technology (CAST). 2018. About Universal Design for Learning. *www.cast.org/our-work/about-udl.html.*

National Research Council. 2005. *How students learn: History, mathematics, and science in the classroom.* Washington, DC: National Academies Press. *https://doi.org/10.17226/10126.*

NGSS Lead States. 2013. *Next Generation Science Standards: For states, by states.* Washington, DC: National Academies Press.

Scruggs, T. E., F. J. Brigham, and M. A. Mastropieri. 2013. Common core science standards: Implications for students with learning disabilities. *Learning Disabilities Research & Practice* 28 (1): 49–57.

# CHAPTER 13

# Reframing Instruction in Teacher Education Settings

**Deborah Hanuscin and Delinda van Garderen**

Perhaps you are an instructional coach, professional development provider, supervising teacher, or teacher educator reading this book, and your intent is to use it as a resource to engage current teachers or preservice teachers in reframing their instruction. As we mentioned at the start of the book, many of the techniques and tools we developed in the Quality Elementary Science Teaching (QuEST) program have become integral to our own work in teacher education. Because of this, we're happy to share an approach we've found that works well.

In particular, we find that "practicing what you preach" is important—that is, using the same frameworks we describe in this book to guide our work with teachers. In the sections that follow, we provide a suggested lesson plan for using the book in different teacher education contexts, such as science methods courses or special education courses; however, this same approach could also be adapted for professional development workshops. We hope that this will be useful to you both as a guide for using the book and as inspiration for developing your own uses for these frameworks.

## Suggested Learning Sequence

In our experience, the kinds of science experiences that future teachers have rarely match the kinds of learning experiences envisioned by current reforms, such as the *Next Generation Science Standards* (*NGSS*; NGSS Lead States

2013). Therefore, an important aspect of our approach in teacher education is to provide students with opportunities to experience science instruction in new ways, so that they can understand these approaches from both a learner perspective and a teacher perspective. In turn, this will support them in applying these frameworks to the design of instruction. You'll notice that we use the 5E Learning Cycle below to organize our lesson! Can you also spot some elements of Universal Design for Learning (UDL)?

## Engage Phase

A first step in *all* teaching should be eliciting students' prior knowledge, and it is no different when teaching teachers. Having our students write a science autobiography (Koch 1990) is a good way to learn more about the experiences they have had in K–12 science, or you could invite them to create podcasts describing a positive and a negative experience with science that can be shared among peers. Nonetheless, these methods don't provide a complete picture of what prospective teachers envision science will be like in their future classrooms. For this reason, we often opt to use a lesson-plan task with our students:

> *You have been invited to teach a science lesson in a local elementary school. You may pick the grade level and topic of your choice. Prepare a lesson plan with enough detail that the classroom teacher will understand what you intend to do. You may assume that all necessary materials are on hand for you to use.*

While this is a fairly open-ended task, we find that it provides an indication of what topics preservice teachers believe should be included in the elementary science curriculum and when it is appropriate to introduce them (something that later can be compared to the *NGSS*). However, you might also choose to assign a particular grade level or science topic and provide more details about the specific classroom of students. By doing so, you would be more likely to elicit preservice teachers' ideas about meeting the needs of diverse learners.

We have students complete this task without consulting any other resources—so that it reflects their own perceptions of what should be taught to elementary school students, how science should be taught, and what should be included in a lesson plan. Reviewing these plans prior to the next step helps us make advance adjustments to our plans, decide who to place in collaborative groups, identify particular questions or issues to raise, and decide how to connect new ideas to what students created for the task. The students will revisit these plans again at the end of the learning sequence, so we typically do not spend much time in class examining them. We find that small-group sharing sessions are sufficient to help students realize that they have ideas that are in some ways similar to and different from those of their peers, and they all are somewhat hesitant about whether what they have prepared is "correct."

## Explore Phase

As a next experience, we immerse students in a model lesson intended to represent *NGSS*-aligned instruction, as well as the 5E instructional framework and principles of UDL. The lesson described in Chapter 11 is one such example that we have used for this purpose. In particular, we find it useful at the onset of the lesson to assign our students one of the learner roles highlighted in that chapter (i.e., Alan, Ainsley, Nick, or Lara), and to challenge them to identify aspects of the lesson they would find more or less difficult as we worked through the experience. We do not have our students actually pretend to be elementary students, as we find they often perceive this to be patronizing; we simply ask them to participate in the lesson and consider how they learn and how that may be similar to or different from how elementary students might experience the lesson.

## Explain Phase

Following the principle of the 5E Learning Cycle (explore before explain), we don't introduce our preservice teachers to the frameworks until they've experienced them in action! After our model lesson, we engage in decomposition of practice (Grossman et al. 2009) in several rounds. This debriefing is preceded by a reading assignment that introduces the frameworks, which are then used to analyze the learning experience. Chapters 1 and 2 of this book could be used as reading assignments before this analysis, as could reading Chapter 11. (However, we recommend discussing students' perceptions of the lesson's alignment with the 5E Learning Cycle and UDL, and unpacking the conceptual storyline of the lesson before this reading, as a way to self-assess their interpretations.)

When decomposing the practice of the 5E Instructional Model, students could create simple graphic organizers to identify the various steps in the learning cycle, the role of the teacher and students in each phase, and the underlying conceptual understandings that are being developed (similar to that in Chapter 2). You might also find it helpful to have students map out the conceptual storyline in a graphic similar to the storyline figure presented in each of the vignette chapters.

When decomposing practices related to UDL, using the learner profiles can be a way to highlight potential barriers and solutions and to connect them to the principles, guidelines, and checkpoints. We also find it useful to have students consider *other* types of learners (perhaps real students whom they encounter in their practicum experiences) and to identify additional barriers and solutions beyond those presented in the lesson. You probably also will have considered potential barriers for your own preservice teachers, and you could decide whether it would be appropriate to share those.

## Extend Phase

To extend students' understanding of the frameworks further, we provide them with an opportunity to put these into practice. Depending on where your course occurs in students' program of study and whether or not they have a practicum experience associated with the course, this might take a variety of forms. The following are suggested activities:

- Analyzing an existing lesson or curriculum for its alignment with the frameworks and making suggestions for improvement
- Analyzing an existing lesson or curriculum with specific learners in mind, to identify barriers and solutions
- Comparing and contrasting two different lessons in terms of their alignment with the frameworks and their potential to meet the needs of diverse learners
- Adapting a lesson or activity (specific to one phase of the 5E or an entire lesson) to align with the frameworks
- Implementing any of the above to evaluate the impact of the suggested improvements on the student learning experience

We also suggest that having students choose another chapter to read and analyze would be a productive activity. Students could "jigsaw" so that small groups read the same vignettes, then present their insights to heterogeneous groups of peers. This would help them consider the ways in which these frameworks apply across different grade levels, topics, and classrooms.

## Evaluate Phase

An important aspect of sense-making is self-evaluation: reflecting on the extent to which one's ideas have changed. At the end of this sequence, we invite students to revisit the lesson they prepared in their lesson-plan task—to examine the extent to which their original lesson reflected the frameworks they have learned about, and how they might now approach the lesson differently. If students wrote a science autobiography, you might also have them revisit their past experiences to better understand what did or did not support their learning from the perspective of the 5E and UDL. Of course, an appropriate summative assessment would also be to have preservice teachers design, teach, and evaluate a lesson in terms of the frameworks featured in this book.

## Closing Thoughts

Learning about these frameworks takes time, as does learning to implement them. Additionally, doing so requires an understanding of the content to be taught and learned. For these reasons, it can seem a daunting task to tackle in a teacher education course, particularly considering all the other topics and skills on which you already focus. The suggested lesson plan that we've provided here requires multiple sessions to complete; however, you may find opportunities to connect to other learning goals that you have for students as well. In particular, the lesson you select may support developing preservice teachers' content knowledge in a particular area, or build their knowledge of students' ideas and how to elicit them, or introduce certain assessment strategies.

The vignettes in this book are based on real teachers' experiences with real students in real classrooms—and, as such, they highlight problems of practice that your own students may encounter. We hope that this provides you with useful role models of teachers who continually work to improve their practice for the benefit of their students.

## References

Grossman, P., C. Compton, D. Igra, M. Ronfeldt, E. Shahan, and P. Williamson. 2009. Teaching practice: A cross-professional perspective. *Teachers College Record* 111 (9): 2055–2100.

Koch, J. 1990. The science autobiography. *Science and Children* 28 (3): 42–43.

NGSS Lead States. 2013. *Next Generation Science Standards: For states, by states.* Washington, DC: National Academies Press.

# Index

Page numbers printed in **boldface type** indicate information contained in tables or figures.

# Index